Principles of polymer morphology

Cambridge Solid State Science Series

EDITORS:
Professor R. W. Cahn
Applied Sciences Laboratory, University of Sussex
Professor M. W. Thompson
Vice-Chancellor, University of East Anglia
Professor I. M. Ward
Department of Physics, University of Leeds

D. C. BASSETT
Reader in Physics, University of Reading

Principles of
polymer morphology

CAMBRIDGE UNIVERSITY PRESS
Cambridge
London   New York   New Rochelle
Melbourne   Sydney

Published by the Press Syndicate of the University of Cambridge
The Pitt Building, Trumpington Street, Cambridge CB2 IRP
32 East 57th Street, New York, NY 10022, USA
296 Beaconsfield Parade, Middle Park, Melbourne 3206, Australia.

First published 1981

Printed in Great Britain at
The Alden Press, Oxford

British Library cataloguing in publication data
Bassett, D C
Principles of polymer morphology.—
(Cambridge solid state science series).
1. Polymers and polymerization
I. Title II. Series
547′.84   QD381   80-49880

ISBN 0 521 23270 8 hard covers
ISBN 0 521 29886 5 paperback

Contents

Preface

Polymer morphology is a young and rapidly growing branch of materials science. Primarily, it is the study of order within macromolecular solids but, in a wider context, it also embraces the processes of formation (crystallization, deformation, etc.) and the consequences for chemical and physical properties. Before 1957 few appreciated that there was significant order worthy of study but the discovery in that year of polymer single crystals and the inference of their chainfolded molecular conformation attracted widespread interest and attention. Much was then achieved rapidly, as Geil's *Polymer Single Crystals* (1963) shows, and development has proceeded apace in the ensuing decades. For the specialist there is a series of excellent reviews but my experience is that postgraduate students find an introduction to these helpful, while for undergraduates a simple summary is essential. This volume is an attempt to provide just such a simple introduction, outlining the essentials of the subject by the use of selected examples. Much new work on representative melt-crystallized morphologies from my own laboratory is also included. Happily, this has made it possible to give a much more comprehensive treatment of lamellar organization than could previously have been contemplated.

The lamella is the continuing theme of the book. It is introduced at the very beginning and, apart from a temporary suspension while the optical textures of spherulites are discussed early in Chapter 2, remains the centre of discussion. The main features of lamellar organization are spread over three chapters. General properties of monolayers, including sectorization and surface structure, come in Chapter 3; other internal subdivisions and three-dimensional development appear in Chapter 4. Then, in Chapter 5, the morphological record is examined for the light it can shed on inherent molecular characteristics of a sample and the treatments to which it has been subjected. Theory is confined to Chapter 6 which has drawn extensively on the review by Hoffman, Davis & Lauritzen (1976) for its approach to the quantitative evaluation of lamellar thickness and growth rate. In Chapter 7 there is discussion of systems giving enhanced chain-extension: the high-pressure (anabaric) crystallization of polyethylene (which turns out to be a particular, though notably instructive, example of melt crystallization) and technologically related processes of strain-induced solidification. Finally, structure–property relationships are touched upon: chemical consequences in Chapter 8 and a selection from morphologically interesting mechanical behaviour in Chapter 9.

Apart from Chapter 6, the treatment throughout is predominantly qualitative. This is a transient stage reflecting the youth of the subject, the establishment of the nature of organization being an essential preliminary before quantitative interpretation can become meaningful. As the basic textures have now either been, or are likely to be, established, the subject can confidently be expected to become rapidly more quantitative in its development.

The illustrations used are primarily of polyethylene and, in lesser degree, other polyolefines. This is not unrelated to these being the examples whose photographs came most readily to hand. (The frequent use of anabaric polyethylenes as exemplars is admittedly idiosyncratic but is, I believe, amply justified by the clarity of interpretation it brings to a wide range of phenomena.) However, concentration on the polyolefines also reflects the fact that these are the most studied and best understood systems. Nor, although highly crystalline, is there any reason to suppose that they are misleading guides to the behaviour of less ordered systems. There are excellent reasons, moreover, why polyolefines should be studied preferentially. Polyethylene, for instance, is the simplest linear macromolecule; it is stable, relatively inert, comparatively easily characterized and exhibits an especially informative and wide range of behaviour.

The list of references at the back of the book refers mainly to the sources of the illustrations; there are few specific citations in the body of the text. This is because I have adopted the same attitude to references that I use when lecturing to undergraduates. I have tended to state what is known, not who did what, unless the persons concerned are associated with substantial portions of the subject. This means that, in general, no credit is given to particular individuals unless, by chance, the illustrations used happen to have come from their paper(s). I believe that, with due passage of time, such treatment becomes inevitable in all subjects and hope that those whose work is not specifically mentioned will be sympathetic. I have, moreover, recommended further reading, usually of reviews, at the end of every chapter and these will be found to be replete with references for those who wish to pursue them.

Assiduous readers may readily detect in the book some reflections of the scientific attitudes of my many friends and sometime colleagues at the University of Bristol, the Bell Telephone Laboratories, the National Bureau of Standards and the University of Reading. The influences of, among others, Sir Charles Frank, A. Keller, H. D. Keith, F. J. Padden,

R. Salovey, J. D. Hoffman and F. Khoury have always been very helpful to me and they are as evident in the text as they have been in my career. No-one other than myself, however, has any responsibility for the book's shortcomings. Many of the above have kindly supplied original prints for the photographs as also have H. Gleiter and J. Peterman, A. J. Kovacs, A. J. Pennings and J. C. Wittman. The largest source of illustrations is, however, my own group at Reading; I am indebted to its various members, and especially Alison Hodge, for their provision. Mrs Margaret Bradley has printed practically all the figures and drawn several drawings. Mrs S. B. Foster typed a difficult handwritten manuscript impeccably while I. M. Ward and the Cambridge University Press exercised considerable patience waiting for its completion. I am also indebted to Professor Ward for helpful comments on the typescript and continual encouragement during its preparation. To all of these people, and others not individually mentioned, I express my warm thanks. Above all, I am profoundly grateful to my wife and children for their ungrudging support throughout, especially when writing has frequently overrun into time which should have been spent with them.

Reading, April, 1980. D. C. Bassett

1   Introduction

1.1   Preliminary survey

Crystalline polymers show ordering at a variety of dimensional levels, from interatomic spacings to macroscopic measures. The study of polymer morphology is primarily concerned to elucidate this organization. The subject is important in its own right as establishing the condition of crystalline solids composed of essentially linear molecules, knowledge of which is a prerequisite for the understanding of certain properties, notably deformation and other mechanical behaviour. In addition, the textures revealed are the products of processes whose mechanisms may also be illuminated by examination of their micro-structural consequences. Furthermore, the morphology is a record of the past history of a sample which, with sufficient understanding, may be read to disclose not only crystallization, annealing or deformation treatments to which it has been subjected but can also provide an indication of certain intrinsic properties, such as the molecular mass range within a specimen or the nature and extent of molecular branch-ing. Evidently, morphological study has a great deal to offer but, despite major advances in the last twenty years, it is fair to state that, with one or two notable exceptions, its full potential has yet to be realized. This applies especially to melt-crystallized polymers and their deformation, i.e. to many commercial products, whose morphologies have hardly been established with any degree of certainty and may still be controver-sial. This situation has, however, begun to change significantly and it seems reasonable to expect that these and other matters will be placed on a much firmer factual basis within the next few years. In that event an appreciation of morphology and its implications will no longer be required only by specialists but will be the proper concern of all polymer materials scientists.

The principal reasons why there are still major areas of ignorance within polymer morphology are technical. Advances have often resulted from the introduction of new techniques and, to that extent, the history of the subject parallels that of corresponding technical advance. At the same time each new item of knowledge is relevant to many other aspects of these complicated materials and leads to a deeper understanding over a wider field. It is scarcely possible, therefore, to treat individual mor-phological topics in isolation. To try to meet this problem, here and again in Chapter 3, a certain amount of basic information is introduced to allow initial discussion. Most, if not all, of this material will, however, need to be reviewed and amplified at appropriate later stages.

Historically, the first detailed knowledge of polymer morphology obtained was that of the crystal structures, i.e. chain packing using the techniques of X-ray crystallography. In addition to the sharp, crystalline reflexions used for the structural analyses, polymer samples generally show diffuse, liquid-like diffraction (Fig. 1.1) indicative of more disordered molecular arrangements whose precise description still largely eludes us today. Such localities are commonly referred to as *amorphous* regions, although the use of descriptions such as 'oriented amorphous areas' in the literature is a recognition of the fact that they are probably not entirely without order, and for this reason the terms *disordered* or *less-ordered* are preferable to amorphous. Nevertheless, the apparent juxtaposition of crystalline and disordered molecular arrangements has given rise to much speculation as to how this might be achieved by the linear thread-like molecules recognized to be characteristic of crystallizable polymers.

For many years the fringed-micelle model held sway. A sketch of a possible arrangement is shown in Fig. 1.2. This contains the idea, already implicit in Fig. 1.1, that, with sufficiently long molecules, these will probably be entangled in part and the regions of crystalline order restricted in size. An individual molecule would be likely to pass through different regions of order and disorder. The concept of two kinds of order (though only a first approximation) has been, and continues to be, very useful in providing the basis of an explanation for the variable densities, melting points, etc., shown by crystallized polymers – and is the reason for their often being described as *semi-crystalline*. The fraction of the material supposed to be fully-crystalline is known as the *(degree of) crystallinity* of a sample and is a widely used parameter in polymer science. This can be a useful concept, provided its use is confined to comparisons of similar textures and provided numerical values assessed by different means are not treated too precisely. In other circumstances, however, an uncritical use of this concept can oversimplify morphological complexities to misleading extents.

The textural scale of the fringed-micelle model was believed, primarily on the basis of crystallite sizes estimated from the widths of X-ray diffraction rings, to be on the scale of a few tens of nm. Such dimensions were not then observable at a time when the first electron microscopes were just being developed. Not until 1945 was it appreciated, first for polyethylene and subsequently for other polymers, that there was additional ordering, on the scale of several μm. This is due to the prevalent

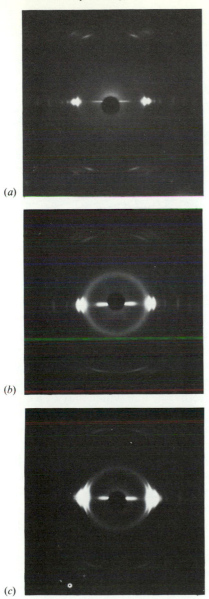

(a)

(b)

(c)

Fig. 1.1. Wide-angle X-ray diffraction patterns of three drawn polyethylenes. In addition to the many sharp and oriented crystalline reflections note the unoriented diffuse or 'amorphous' ring.

This is much stronger in the very high molecular mass polymer (b) and in branched material (c) than in the typical commercial linear polymer (a).

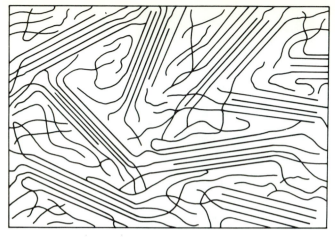

Fig. 1.2. The fringed-micelle model of polymeric texture. (After Bunn, 1953.)

crystallization of high polymers as spherulites, literally little spheres, a mode previously found mostly in viscous minerals.

The study of spherulites with the polarizing optical microscope then became the second major area of polymer morphology to be investigated. At this stage the link of morphology with properties becomes particularly evident. On one hand spherulites as optical inhomogeneities on a scale of μm scatter light strongly and are responsible for the cloudiness of, for example, polyethylene film which may be a disadvantage for a packaging material, on the other spherulites are associated with, and believed to be due to, the segregation of different molecular species in a sample. For example shorter molecules are likely to predominate in inter-spherulitic boundaries and give these regions different mechanical properties leading in certain circumstances to preferential fracture between spherulites. When such behaviour results it becomes particularly important to understand what the morphological texture is, how it formed, how it can be controlled and, if possible, modified to give improved properties.

It was by no means obvious how the supposed fringed-micellar arrangement of molecules at the 20 nm level fitted within spherulites two orders of magnitude larger in scale, particularly in view of the fact that the molecular chain is generally tangential to spherulites instead of radial as had been expected. Matters became a good deal clearer when the introduction of the first generation of modern electron microscopes, coinciding with the synthesis of highly linear and stereoregular polymers following Ziegler and Natta, led to the discovery of individual polymer

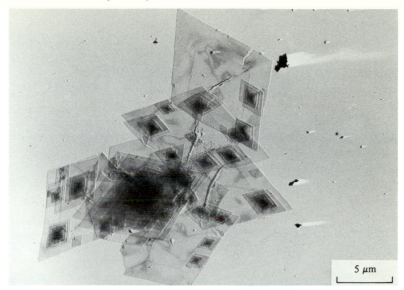

5 µm

Fig. 1.3. Solution-grown lamellae of polyethylene. Among their many features are an internal step and change of habit due to a fall in crystallization temperature, evidence of sectors and non-planarity and the development of growth spirals as planar growth surfaces become unstable.

crystals grown from very dilute solutions (Fig. 1.3). The very idea of polymer molecules forming separate crystals was alien to the fringed-micelle model, yet it has been shown that, for example, a whole sample of linear polyethylene can be precipitated as crystals of dimensions say 12 nm thick by 10 µm wide bearing a remarkable resemblance to those of the aliphatic *n*-paraffins (which are polyethylenes of very low molecular weight). This similarity suggested the even more remarkable fact, correctly deduced by Keller, that molecules – typically 5–10 µm long – lay in the crystals with their lengths across the thin dimension (12 nm) of the lamellae. The inescapable conclusion was that the chains must fold back on themselves repetitively at each crystal surface alternately, a phenomenon now known to be widespread and called *chainfolding* (Fig. 1.4). These twin discoveries of polymer lamellar crystals and chainfolding lie at the heart of modern understanding of polymeric morphologies and have also stimulated enormous interest and activity in the subject.

With samples consisting entirely of individual crystals and a variety of morphological evidence indicative of ordered folding there was a period when it was difficult to locate the 'amorphous' regions associated with crystalline polymers. It soon transpired, however, that these were to be

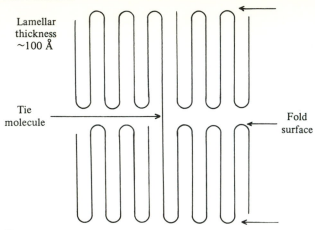

Lamellar
thickness
~100 Å

Tie
molecule

Fold
surface

Fig. 1.4. A chainfolded conformation (schematic). (From Rees & Bassett, 1971.)

identified largely with regions at or between the *fold surfaces* (i.e. the large basal surfaces where molecules turn back on themselves). These have significantly different properties from lamellar interiors. For example, although the density of the crystallographic subcell of polyethylene is 1.00 g cm^{-3} at 20 °C, that of typical individual crystals is about 0.98 g cm^{-3}, increasing with their thickness. This has led us back to a two-phase model of texture, with crystalline lamellar interiors sandwiching disordered surface and interlamellar regions, to a first approximation. Particularly when mechanical properties are considered, this needs to be extended by considering lateral discontinuities between lamellae, etc. Nevertheless, the recognition of lamellae with their surface regions acting as mechanical discontinuities or channels for the entry of reagents has greatly improved our understanding of both physical and chemical behaviour of crystalline polymers.

So far as bulk, melt-crystallized polymers are concerned, evidence for their containing lamellae has been much harder to obtain than for solution-grown specimens but it has always been clear that folding would be neither so complete nor so regular as from solution. Nevertheless, evidence for lamellae has accumulated mostly by comparison with the similar behaviour of solution-grown crystals using a variety of techniques such as small-angle X-ray scattering, thermal measurements and molecular mass measurements of chemically degraded materials, as instrumental facilities have become available. What has not been possible, until very recently, is to observe representative lamellae microscopically in melt-crystallized polymer. The introduction of new techniques of specimen preparation has now made this possible and there is little doubt that it will lead to much more firmly based understanding of the organization and crystallization of polymers grown from the melt. At the same time the new tool of small-angle neutron scattering has begun to give information on the conformations of individual molecules in both molten and melt-crystallized polymers. At first sight these appear to be very similar so that it has begun to be suggested that regular chainfolding, and possibly chainfolding of any form, cannot be prevalent in melt-crystallized samples despite their lamellar texture. This is naturally highly controversial but it is also an indication that the techniques available are at long last probably sufficient to allow us to begin to characterize, at the lamellar level and beyond, the complex morphologies of melt-crystallized polymers. In subsequent chapters the outlines of such a characterization and its consequences will be attempted.

1.2   Molecular and crystal structures

Excellent detailed reviews of these topics and their interrelation are available (e.g., Geil, 1963, Keith, 1963, and Kitaigorodskii, 1961) and, for polymers, this is a particularly well-understood area. In this volume only the main principles involved will be outlined.

Practically all monomeric and oligomeric substances are able to crystallize; whether or not polymers of these same monomers can also be crystalline depends primarily on the regularity (chemical, geometrical and spatial) of the macromolecular chains. In general terms, the more regular the polymer the more likely it is to crystallize. The need for chemical regularity is fundamental. Polyethylene is a good illustration of the sensitivity to this factor in that low-density polymers with say 30 short-chain branches randomly distributed per 1000 carbon atoms of the chain, melt some 20 K below high-density materials with only about three such defects per 1000 carbon atoms. This is predominantly due to the poorer quality of crystalline order in the first case compared with the second. In most polymers, monomer units must be arranged in the same head-to-tail sequence else, as in polymethylmethacrylate, the condensed polymer will not be crystalline. (A notable exception to this is polyvinyl-alcohol $[-CH_2\ CHOH-]_n$ which is insensitive to the reversal of hydroxyl and hydrogen positions presumably because of their similar sizes.) Even with perfect chemical and geometrical repetition along it, however, a macromolecular chain must also be spatially regular if neighbours are to be able to fit well together. A well-known example is that of the two isomers of polyisoprene

$$-[CH_2-C(CH_3) = CH-CH_2]_n-$$

Both cis and trans configurations are possible with respect to the double bond. The latter results in a much-less-puckered chain, giving the highly crystalline guttapercha whereas the former is the basis of natural rubber whose crystallization is difficult and is greatly enhanced by the imposition of tensile stress. A second familiar case is provided by the polyolefines $-[CH_2-CHR]_n-$ where the substituent R groups have to be systematically arranged along the chain to permit crystallization. There are two spatially ordered arrangements, *isotactic* and *syndiotactic*, in which, as Fig. 1.5 shows, the pendant groups have, respectively, the

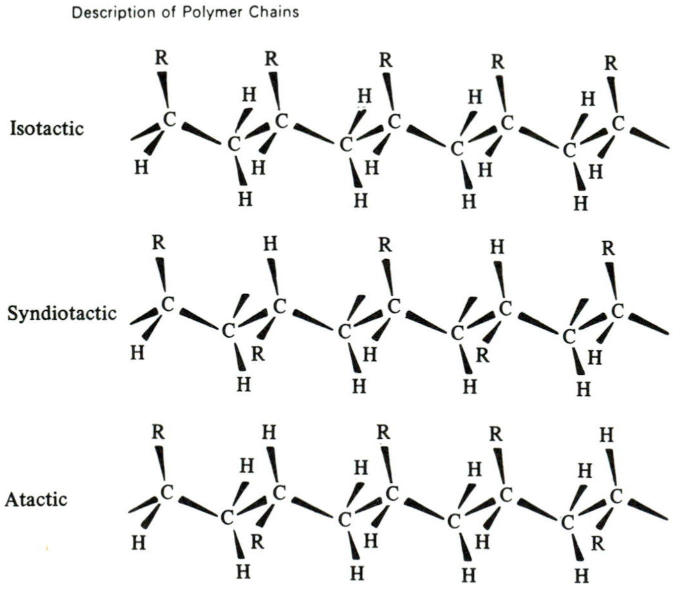

Description of Polymer Chains

Fig. 1.5. Chain configurations in poly (α olefines). (From Schultz, 1974.)

same placement or an alternating one in relation to atoms in the main chain. Disordered sequences are known as *atactic*. It is a commonplace that whereas atactic polystyrene and polypropylene cannot crystallize, their isotactic counterparts will do so.

The chain configurations of an all-carbon backbone are based upon the tetrahedral valence bonds of the carbon atom. For simple aliphatic compounds the diagram of potential energy against angle of rotation around a single covalent bond is as shown in Fig. 1.6. Zero rotation corresponds to a trans configuration, T, between next nearest neighbour bonds along the molecule. The all-trans configuration is adopted by the crystallized polyethylene molecule both in the usual orthorhombic structure (Fig. 1.7) and in the less common monoclinic (triclinic) modification found in deformed samples (Fig. 9.8). A different form is found at high pressures (~ 0.5 GPa), in which gauche bonds, G or $\bar{G}$, are incorporated into the chain, probably because this lowers the free energy by increasing the entropy of the system. This particular structure has still to be fully elucidated but it has been proposed that it consists of alternating regions of all-trans bonds interrupted by TGT$\bar{G}$ sequences. Such sequences are helical and helical molecular conformations are widespread among crystalline synthetic polymers. Strictly, the all-trans configuration is a 2_1 helix, i.e. two repeat units (CH_2 groups) per 360° turn of the helix. In the poly (α olefines)—[CH_2—CHR]$_n$—it is possible for the side groups R to pack much more economically if the molecule is helical rather than in an all-trans configuration. For an all-trans isotactic polymer, for example, all the side groups would lie adjacent on one side of the molecule, whereas a helical conformation allows them to be spread more suitably round the chain. Superposed on the basic bond sequence, for example all-trans for polyethylene, alternately trans–gauche for isotactic polypropylene, may be distortions to give the optimum packing. A good example is PTFE which forms a 13_6 helix in the form stable below 19 °C and a 15_7 helix above 19 °C. Both these

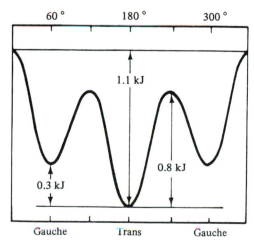

Fig. 1.6. Potential energy as a function of bond rotation for a covalent carbon–carbon bond, illustrated for ethylene chloride. (After Schultz, 1974.)

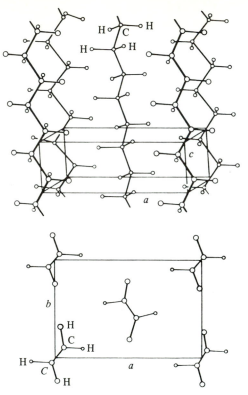

Fig. 1.7. The orthorhombic polyethylene structure. (From Bunn, 1953.)

helices can be regarded as distorted all-trans configurations (i.e. slightly twisted 2_1 helices).

To the extent that molecular chains can be regarded as rigid rods with the van der Waals' bonding between molecules weaker than intramolecular covalent linkages, they will tend to pack together in a simple hexagonal array, i.e. with sixfold coordination. Certain structures are, indeed, truly hexagonal; for example, polyoxymethylene, hexagonal polypropylene and the disordered high-pressure phase of polyethylene. More often, however, lower symmetries prevail, such as in orthorhombic polyethylene (Fig. 1.7) or monoclinic polypropylene (Fig. 1.8) which are distorted hexagonal structures. Other factors can also intervene. The phenyl groups of polystyrene place the isotactic molecule in a 3_1 helix whose circumscribed triangular prism brings threefold coordination. In a similar way it is the provision of optimum packing of the side groups to which the tetragonal structure of isotactic poly(4–methylpentene–1) is

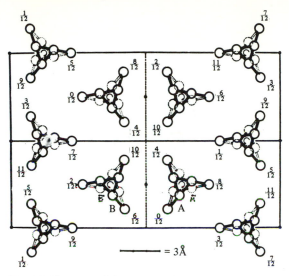

Fig. 1.8. The monoclinic polypropylene structure, projected on {001}. (From Natta & Corradini, 1960.)

attributed. Sometimes constitutent groups cause a chain to adopt a more planar than rod-like shape with consequences for the crystal structure. Thus the all-trans planar zigzags of the n-paraffins and polyethylene can pack together either in the distorted hexagonal structure of the ortho-rhombic form (Fig. 1.7) or they may adopt the slightly higher energy monoclinic subcell in which they are all parallel. Other examples are polyisoprene, with its cis configuration, and polyethylene terephthalate (Fig. 1.9), the latter deriving its planarity from the phenyl groups in the main chain, which all lie parallel.

Lastly, there are the structural developments associated with hydro-gen bonding and the polypeptide linkage

$$-\mathrm{C-NH-}$$
$$\underset{\mathrm{O}}{\|}$$

which are so important in biopolymers. Among synthetic macromole-cules these are associated particularly with the nylons. Hydrogen bond-ing is a relatively strong linkage which, when it occurs between adjacent stems, tends to form them into sheets, with weaker external than internal binding; three examples are shown in Fig. 1.10.

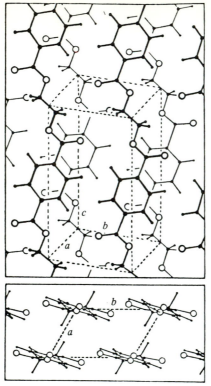

Fig. 1.9. Molecular packing in polyethylene terephthalate. (From Daubeny, Bunn & Brown, 1954.)

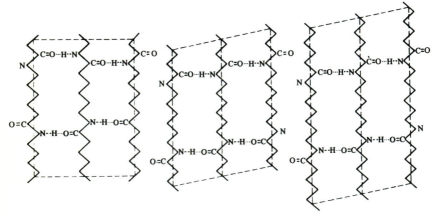

Fig. 1.10. Hydrogen bonded sheets of molecules in, from left to right, nylons 6, 66 and 610. The dotted lines outline the faces of the unit cells. (From Keith, 1963.)

1.3   Methodology

Examples have already been given of the way the introduction of new
techniques has continually led to advances in the understanding of
polymer morphology. No two methods reveal the same feature of
texture in precisely the same way and the synthesis of a reasonably
comprehensive picture requires information on as many aspects as
possible from as many means as are available. The use of different
techniques to complement each other is essential in morphological
enquiry and is likely to involve more and more methods as the sophisti-
cation of textural characterization increases. Until now, with the subject
still in its infancy, there is little doubt that microscopy in its various
guises has been the most valuable tool. It possesses the overriding
advantage of revealing textures and their distribution in direct space. It
also allows the eye to assess the situation directly, with the considerable
gain in information that that implies. Above all, it is microscopy which
reveals the nature of the textural elements, i.e. the type of model which
can then be used in the interpretation of results produced by other
techniques. It has been said, for instance, that polymer lamellae would
never have been inferred had they not been discovered with the microscope. It is often advantageous, nevertheless, to use other methods,
notably those involving diffraction or reciprocal space, when measurement or details of molecular orientation are required.

There are two reasons for this. Firstly, microscopy examines only a
small part of a specimen at a time, increasingly so the higher the
magnification, with the result that the problems of deciding what is a
representative morphology and the dangers of selecting special features
both increase with the enlargement of the image. Secondly, what is
usually wanted is a parameter averaged over the whole sample. Microscopy can only provide this by statistical sampling procedures whereas
other methods may give the result straightaway. Consequently, the
measurement of average lamellar thickness, d, is usually made with
low-angle X-ray scattering which gives maxima at diffraction angles 2θ
related, in the simplest cases, by

$$\lambda = 2d\theta$$

for wavelength λ. This applies, for example, to aggregates of solution-
grown lamellae but with melt-crystallized specimens the situation is
more complicated with often only one or two diffraction maxima
observable. Such information does not allow a unique interpretation of
the diffracting structure and one must proceed by comparison with
predictions of models of lamellar texture based, ultimately, on microscopic observations.

For lamellae thicker than ~ 100 nm low-angle X-ray diffraction becomes technically more difficult. Partly for this reason, and partly because the spread of thicknesses may be too great to produce well-defined diffraction maxima, one is then likely to have to revert to microscopy and statistical sampling to measure lamellar thicknesses in this range. Such was the case for polyethylene crystallized anabarically, for example, at 5 kbar. More recently, though, it has been demonstrated that the distribution and average values of crystal thickness in such samples can be measured by what is termed the nitration/GPC technique. For this, polyethylene is digested with fuming nitric acid (typically for three days at $60°C$) which, to a first approximation, removes all original lamellar surfaces thereby cutting all molecules into segments of a length equal to the lamellar thickness measured along c. These values may be determined by gel permeation chromatography, GPC, of the degraded product.

Application of nitration/GPC to solution-grown polyethylene lamellae, as was introduced at Bristol (Williams, Blundell, Keller & Ward, 1968), can yield still further information, viz. the number of carbon atoms involved in the molecular chain folding back on itself. This is done by noting that, at least for solution-grown crystals, the molecular lengths left after nitration usually contain proportions of integral traverses of the crystal corresponding to folds which were not severed by the nitric acid. The difference between two single traverses and one double, or one double and one treble, will thus be the amount (in the folding region) removed in the process. This is one non-microscopic method, therefore, of gaining insight into the extent and nature of the surface regions of lamellae and will be discussed subsequently in Chapter 3. Alternatively, one can attempt to measure the fully-extended (stem) length of a molecule inside a crystal from the frequency of the longitudinal acoustic mode using Raman spectroscopy. It is synthesis of all such results which builds up the best available picture of polymer morphology.

Conversely, experiments which fall significantly behind modern practice in morphological characterization are liable to be of limited value because they must tend neither to illuminate, nor to be illuminated by, the current understanding of the subject. A few years ago it was not uncommon to characterize a polymer only by its name, density (possibly leading to a figure for crystallinity) and a description such as quenched or slowly cooled. Nowadays this is most probably far too little information. One would now wish to know also such quantities as the mass distribution, tacticity and/or degree of branching of the molecules, the melting endotherm, the optical texture, the lamellar thickness(es), structures revealed in an electron microscope, the wide-angle X-ray pattern (to reveal orientation and/or the presence of additional crystal structures) and, particularly for deformed samples, the small-angle X-ray pattern as well as a detailed processing history. This, and more, is commonly done in the specialist morphological laboratories. Because of the basic importance of morphology in many aspects of polymeric behaviour, it is likely to become the required practice in other laboratories concerned with polymeric materials science as the interrelationships of microstructure and properties become better understood.

1.4   Further reading

A number of excellent articles exists on the related crystallography of organic and macromolecular solids. The underlying organizational principles are expounded in Kitaigorodskii (1961) while Keith (1963) discusses detailed relationships between molecular and crystal structure. Extensive compilations of data are to be found in Geil (1963) and in Tadokoro (1979), the latter being a specialist text concerned with all aspects of polymeric crystallography.

A selection of reviews of polymer morphology may also be helpful, particularly in relation to the first five chapters. In addition to Geil (1963) and Keith (1963) those by Keller (1958, 1968) and by Khoury & Passaglia (1976) are recommended. All are replete with references and, with their relative dates, reflect the historical development of the subject.

2 Spherulites

The aspect of polymer morphology most likely to be met with, because it is so widespread and can be seen in an optical microscope, is spherulitic ordering, i.e. spherulites in various stages of development. Spherulites, literally little spheres, with diameters usually in the range 0.5–100 μm are characteristic of polymers which have crystallized from the melt in the absence of significant stress or flow. They are not, however, a specifically polymeric phenomenon being found also in minerals and simple substances grown from melts which were viscous and impure. One believes, therefore, that spherulites form as a consequence of certain general crystal growth conditions which happen to pertain to polymers. For these reasons, and to emphasize the essential independence of spherulitic textures from the lamellar units built into them, spherulites are discussed before consideration of polymeric lamellae and their formation. The treatment given follows rather closely that of Keith & Padden (1963), which relates textural variations in scale to a diffusion coefficient and the growth rate.

2.1 Spherulitic morphologies

The three essential minimum attributes of spherulites may be inferred from their appearance in Fig. 2.1, illustrating specimens of two polymers and also malonamide, viewed between crossed polars in an optical microscope. Fig. 2.1a is of what may be called a classic spherulite of isotactic polystyrene. The bright contrast derives from birefringence and indicates a crystalline entity. Superposed on this is the black Maltese cross with arms lying parallel to the extinction directions of polarizer and analyser. When the specimen is rotated in its own plane the cross remains stationary. This implies, as will be explained in the next section, that all radii are crystallographically equivalent to within the resolution employed. Accordingly, spherulites are polycrystalline with equivalent radial units. This description fits the object of Fig. 2.1b which, however, unlike the spherulites of Fig. 2.1a, fails to fill space; the radial fibrils simply diverge. Only if they branch, creating more fibrils at greater radial distances, will space be uniformly filled. Fig. 2.1c indicates how this can happen, although this particular object, which has a very open texture, probably branches at more regular angles than is typical. The

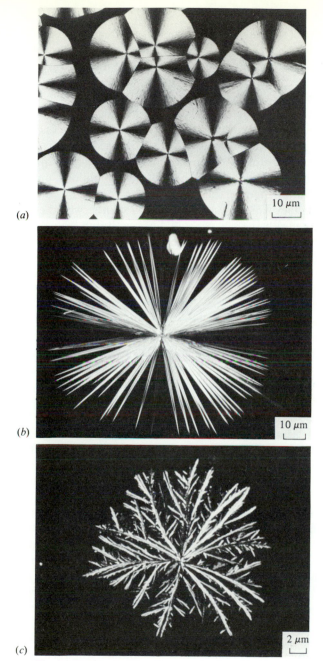

Fig. 2.1. (a) Classic spherulites growing in a thin film of isotactic polystyrene. Crossed polars. (From Keith, 1963.) (b) Radiating growth of malonamide from a melt containing 10% *d*-tartaric acid. Crossed polars. (From Keith & Padden, 1963.) (c) Branching fibrils in a spherulite of isotactic polypropylene grown from a melt diluted with 90% atactic polymer. Crossed polars. (From Keith & Padden, 1964a.)

microstructure of spherulites is thus a polycrystalline array of equivalent radiating, branching units tending to produce a spherical envelope.

This fine structure will also serve as a means of identifying a spherulitic entity because, paradoxically, the spherical outline is not an essential requirement but represents only a transient stage in growth. Precisely how a spherulite develops depends, in the first instance, on how it is nucleated. A common, but again not an essential, progression is that depicted in Fig. 2.2†, beginning with a fibre and evolving through sheaf-like embryos before attaining a spherical envelope. This is, however, only maintained, even under thermostatic conditions, until neighbouring spherulites impinge; then they become polyhedral. If nucleated simultaneously they will share a planar interface, otherwise the common boundary will, ideally, be a hyperboloid of revolution (Fig. 2.1a). This succession of shapes for the same object growing isothermally makes it preferable to define the nature of a spherulite not in terms of whatever outer profile it happens to have formed but rather on the basis of the constant nature of its fine structure. This we shall do and simply refer to different stages in development as immature or mature.

Mature spherulites have the same crystallographic axis along every radius. In this they differ from dendrites, which are also highly branched solids but possess a common single crystal orientation. Even this distinction should not be pressed too far because the two have no clear divide but merge continuously as growth conditions vary.

In the investigation of spherulitic morphologies, polarizing microscopy has been the predominant and most informative tool. Very often samples have been crystallized in thin films to facilitate observation during growth, as in Fig. 2.1. This tends to produce greater symmetry of form since the crystalline circles which result resemble rather closely diametral sections (i.e. through the centre) of spherulites grown in bulk (unrestricted volume). (To examine bulk samples similarly one has little alternative but to cut sections.) Much effort has been expended in adjusting conditions to open up structures for readier examination (e.g. Fig. 2.1c). Even so it is evident that details of the radiating units often lie tantalizingly just below the resolution of the optical microscope. One might imagine that all one has to do to observe these structures is just to cut suitably thin sections of spherulites and examine them with the greater resolution of the electron microscope. This, however, has proved a frustrating and largely fruitless exercise.

It is not trivial to cut sections, say 100 nm thick, without causing severe distortion but, by cooling polymers and making them brittle, it has been done. The fundamental difficulties arise rather with the micros-

† This figure is taken from the Ford Prize address (1973) by H. D. Keith and F. J. Padden.

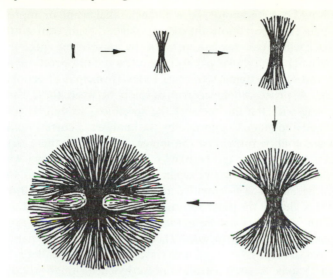

Fig. 2.2. A common progression of habits leading to the spherical form of a mature spherulite.

copy because of radiation damage. This is always a great handicap in the examination of carbon polymers and causes rapid loss of crystallinity whenever specimens are exposed to the electron beam. Thin sections, however, appear especially susceptible. They also exhibit radiation-induced contrast caused by internal mass transport which smears out structural detail. In consequence, electron microscopy of melt-crystallized polymers has proved rather unprofitable until very recently. The change is due to two new complementary techniques, chlorosulphonation of polyethylene and permanganic etching of various polyolefines, of which details are given in Chapter 4. With them it has at last proved possible to characterize the lamellar morphology of one comparatively simple, spherulitic system, viz. polyethylene crystallized at high temperatures. While this is an auspicious development, the body of our knowledge and understanding of spherulites still rests on optical observations. It is to these, therefore, that this chapter must largely be confined.

2.1.1   Molecular orientation

Observation between crossed polars is the basis of standard mineralogical methods for determining the orientation of optically anisotropic (i.e. non-cubic) crystals. Similar information is available for spherulites by the observation of extinction, i.e. blackness in the image. The illumination should be orthoscopic, i.e. a plane parallel beam, or approximately so. Under these conditions the plane polarized beam transmitted by the polarizer will travel through an optically anisotropic specimen as two orthogonal beams whose vibration (electric vector) directions are parallel to those of extreme refractive index (principal directions) in the section. A phase difference will develop between these two beams, increasing with the thickness of the specimen, so that when they are combined and made to interfere by the analyser, a corresponding interference colour is formed in the image. Only when there is no phase difference will the image be black. This can arise in two ways. *Zero amplitude extinction* occurs when only one beam travels through a crystal because it is oriented so that one of its principal directions is parallel to that of the polarizer; the other will be parallel to the analyser direction. *Zero birefringence extinction*, on the other hand, is due to the disappearance of birefringence when the illumination is parallel to an optic axis of the crystal. For a uniaxial crystal this is parallel to the major symmetry axis. As an illustration polyethylene (though strictly biaxial) is almost uniaxial, with the refractive index parallel to the molecular chain substantially greater than along any perpendicular axis, so that zero birefringence extinction occurs in those regions where the molecular chain is parallel to the illumination.

One can distinguish between zero amplitude and zero birefringence extinction by the simple expedient of inserting crossed quarter-wave plates on either side of the specimen. The first of these introduces a 90 ° phase difference which the second removes. The blackness of zero birefringence extinction remains because the specimen itself can create no phase difference. However, for zero amplitude extinction, where previously only one beam passed through the crystal, now there will be two so that the inherent colour due to the crystal will be seen instead of blackness. The Maltese cross, for example, disappears under these circumstances (Fig. 2.3).

The Maltese cross occurs when the principal optical directions of the radial units fall parallel to those of the polarizing microscope. The significance of the cross remaining stationary while a specimen rotates is then that each radial unit has the same extinction directions. This, in turn, signifies that the major refractive index is then parallel to one of the polars.† A typical Maltese cross, oriented parallel to the polars as in Fig. 2.1*a*, thus means that the major refractive index (in polyethylene the chain direction) is either parallel or perpendicular to a radius. In fact the

---

† The possibility of the projection of the major refractive index being parallel to one of the polars is removed if the cross appears the same for all sections, or in all spherulites, because one is then finding all rotations about a radius to be equivalent.

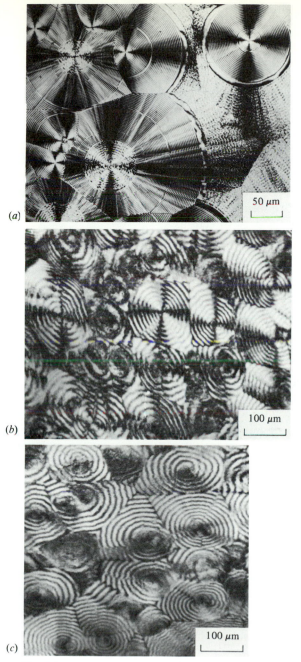

Fig. 2.3. (a) Banded spherulites grown in a thin film of poly(trimethylene glutarate). Note the two different ring spacings due to growth at two temperatures. Crossed polars. (From Keller, 1959b.) (b) Zigzag Maltese crosses in banded polyethylene spherulites. Crossed polars. (c) Zero-birefringence extinction in banded polyethylene spherulites. Crossed polars and crossed quarter-wave plates.

polyethylene chain is perpendicular to the spherulite radius, a feature which is typical of polymer spherulites.

This, too, can be established optically from the sign of birefringence. Insertion of a first order red plate, also known as a sensitive tint plate, between the crossed polars yields a blue colour for slightly increased birefringence, i.e. when the direction of greatest refractive index is within 45° of that of the plate, and yellow otherwise. A section of suitable thickness will thus show its spherulites with yellow and blue quarters. The relative positions of the colours in relation to the tint plate reveal the orientation of the refractive index ellipsoid and with it, in simple cases, the molecular chain axis. This deduction may be simply checked by comparison with a section of drawn material whose molecules will tend to be parallel to the draw direction.

In certain cases these orientations have also been precisely deduced by X-ray microdiffraction, in which areas as small as ~10 μm have been selected and photographed through a capillary made from lead-glass or other suitable material placed in contact with the specimen. In this way the radial direction in polyethylene spherulites has always been found to be the crystallographic b axis. The once surprising orientation of the molecules perpendicular to a radius, rather than parallel as had been expected, is now seen as a natural consequence of there being chain-folded lamellae growing outwards parallel to radii.

2.1.2   Banded spherulites

In Fig. 2.3 are examples of more complicated spherulites which, in addition to the Maltese cross, also show concentric extinction rings. These are a zero birefringence phenomenon (Fig. 2.3c) indicating that at the location of the rings the molecular axis (c) is parallel to the illumination. In fact tilting experiments show that the chain axis spirals round the radial direction. It may also be noted that in Fig. 2.3b, which is of a thin section cut from the bulk, the Maltese cross has degenerated into a zigzag connecting the bands.† This is a consequence of the radial units being inclined to the plane of a non-diametral section. A similar effect can also occur in spherulites grown in thin films when, as a result of thermal gradients or other cause, growth is directed from one surface to another at some angle to the film plane. Zigzags of rather larger amplitude may also occur in simpler, unbanded spherulites when, to quote

† Fig. 2.3b shows a variation in intensity round the rings due to residual birefringence of order a quarter of a wavelength in the microscope itself. This is a common effect but needs to be removed by compensation (with a quarter-wave plate) before, for example, uniform ring patterns as in Fig. 2.3c can be obtained.

Keith (1963), 'the crystallographic directions of the twisted fibrils are such that optic axes are prevented from falling in tangential directions'.

On an optical level all these extinction effects are understood; the reader is referred to the original papers of Keith & Padden (1959*a,b*), of Keller (1959*b*) and of Price (1959) for further details. At a more fundamental level, however, there is no accepted explanation. It is believed that the cause of the twisted crystallization is stresses set up during crystallization, but as to what these might be in particular there is no agreement. Facts which are known to apply generally are that the ring spacing (pitch of the twist) increases with crystallization temperature (Fig. 2.3*a*) and may disappear altogether for crystallization at low supercoolings. Also in any one region the twist is uniform and of the same hand – as may be demonstrated by rotating samples about a radius and observing the movement of fringes – but it is common to find radial boundaries at which there is a change of hand, but not of period. Progress in understanding this phenomenon is much hindered by lack of detailed knowledge of spherulitic microstructure and it seems that advance on this topic may be a prerequisite of understanding twisted crystallization.

2.1.3   Nucleation and growth

The polarizing microscope has also been much used to study the growth rates of spherulites, i.e. the velocity of radial advance. All studies show, as in Fig. 2.4, that this quantity is remarkably linear with time at a given

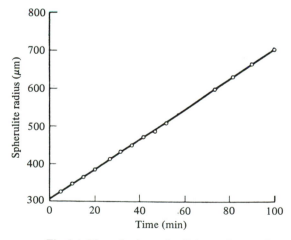

Fig. 2.4. Linear isothermal radial growth rate of a spherulite growing from a blend of 20% isotactic and 80% atactic polypropylene. (From Keith & Padden, 1964*b*.)

temperature, except when spherulites near impingement or when the viscosity of the melt is deliberately reduced. This is, in fact, a crucial point in the understanding of spherulitic growth. The variation with temperature typically takes the form of Fig. 2.5 whose detailed interpretation is considered in Chapter 6. It has long been recognized, however, that the characteristic shape is a consequence of growth being slowed by increasing viscosity at lower temperatures and by diminishing thermodynamic drive (i.e. supercooling) as the melting point is approached. For polyethylene, growth is so fast that only the higher end of this curve can be realized while for other polymers such as isotactic polypropylene and isotactic polystyrene the whole range can be measured.

The extent over which growth rate can be measured depends on spherulite size and this in turn depends on the relation between nucleation rate and growth rate. In simple terms, if nucleation is slow and growth fast by comparison, a few large spherulites will result. Conversely, rapid and considerable nucleation will lead to a profusion of spherulites which may well be immature. Polyethylene often behaves in this way resulting in a texture of very many small embryos which may be barely recognizable as such. If spherulite size is to be controlled, and this can be a desirable end in just the same way as control of grain size is

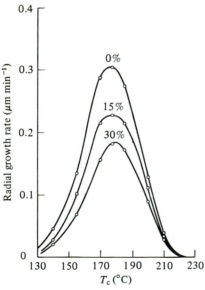

Fig. 2.5. Variation of radial growth rates with crystallization temperature in isotactic polystyrene and blends containing 15% and 30% of atactic polymer. (From Keith & Padden, 1964*b*.)

important in metallurgy, then it is through varying nucleation rates that this is best attempted.

This is possible because, almost always, nucleation in polymers is *heterogeneous* in origin, i.e. it is initiated on non-polymeric particles. Were it *homogeneous*, forming on the polymer itself, many fewer possibilities would present themselves. One could, for example, alter the molecular mass but this will also affect other properties and may not be desirable. That nucleation is generally heterogeneous is demonstrable in two ways. Firstly, if a polymer is melted and recrystallized then spherulites will often reappear at their original sites. It is usually supposed that this is due to nucleation in cracks of, for example, foreign particles. One can then explain why, if the melt is taken to successively higher temperatures before recrystallization, the number of nuclei decreases and why to destroy previous nucleation sites it is necessary to hold the polymer for considerable times at temperatures well above its melting point. A second experiment of considerable importance is the droplet experiment whereby a sample is dispersed into separate microspheres say 20 μm in diameter prior to crystallization. It is then found that a small proportion of droplets will crystallize at the same temperature as for an undivided sample but the majority remain molten to substantially lower temperatures (80 °C instead of 125 °C for polyethylene). The explanation is that the few heterogeneous nuclei have been isolated in their own droplets and rendered impotent so far as the remainder of the sample is concerned, which subsequently crystallizes homogeneously. In a bulk sample, however, once heterogeneous nucleation has started growth will proceed through the entire sample. This is the basis for adding heterogeneities deliberately to samples to control spherulite size and development.

No matter whether nucleation is heterogeneous or homogeneous it is still strongly affected by molecular weight, with longer molecules usually initiating crystallization of the polymer. Presumably this is because the longer a molecule is, the greater the chance of being able to adopt a suitable conformation. As a consequence, if it is desired to grow large spherulites of, for example, polyethylene or poly(4–methylpentene–1), one has only to heat the polymer sufficiently to cause slight degradation. The loss of the longer molecules (and also the creation of additional very short ones, as will presently be explained) both help to grow fewer and more open spherulitic structures. In this way one can readily observe the early stages of spherulitic growth. Fig. 2.6 shows a sheaf-like stage, somewhat similar to that sketched in Fig. 2.2. Other initial sequences can also be observed, notably those developing from apparent single crystals in a manner similar to that of Fig. 2.11. The variety of early

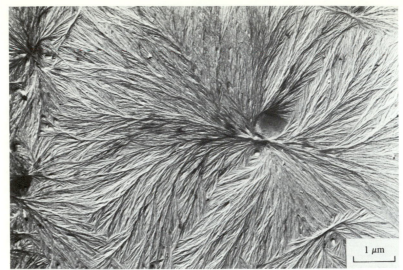

1 μm

Fig. 2.6. Spherulitic sheaf in a melt-crystallized film of poly(4–methylpen-tene–1).

stages which exists makes the point that it is not the origin of a spherulite which is important to its final shape but the way in which growth proceeds.

It is convenient at this stage to illustrate two common modifications of spherulitic texture often found in commercial material. These are shown in Fig. 2.7. Both result from massive nucleation, in Fig 2.7a along lines due to flow within the material, and in Fig. 2.7b to cooling of the external surface such as could occur during extrusion. With so many nuclei, growth can only proceed perpendicular to the original line or sheet forming what are known as *row structures* and *transcrystalline layers* respectively. In both cases the transverse growth is equivalent to that along the radius of a spherulite, possessing the same crystallographic axis. Related effects have been observed in polymer blends (Fig. 2.8) where nucleation can be produced on interphase boundaries leading at least to modified kinetics and, presumably, modified properties.

The size of spherulites can obviously be measured from thin sections with suitable sampling techniques, but a more rapid technique giving an average which is widely used is due to Stein & Rhodes (1960) and is known as low-angle light scattering. The method is essentially a diffraction experiment from assemblies of spherulites using the anisotropy of response to a polarized light beam, with two main variants according to the state of polarization used. For details the reader is referred to the original publication.

(a)

(b)

Fig. 2.7. (a) Row structures, with diameters of tens of μm, in isotactic polypropylene. (From Maxwell, 1965.) (b) A transcrystalline layer growing inwards from the surface of isotactic polypropylene in contact with PTFE. (From Fitchmun & Newman, 1970.)

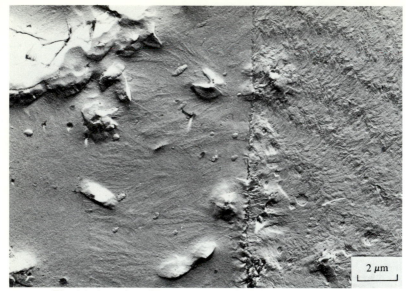

Fig. 2.8. Interphase boundary in a polyethylene and polypropylene blend. Nucleation of both polymers has occurred on their common surface, profusely for polypropylene on the left and initiating a banded polyethylene spherulite on the right. (From Olley, Hodge & Bassett, 1979.)

2.2   The theory of Keith and Padden

There are two major features common to all the variants of spherulitic form. The first is the microstructure, in which every radial element is equivalent in mature spherulites, and the second that the isothermal growth rate is constant and independent of radius. The theory of Keith & Padden (1963) presents a semiquantitative rationale of these properties in terms of a general crystal growth mechanism. Other features such as banding, or sheaf-like precursors, are not essential elements although the latter is easily understood in terms of the theory; the former is, as we have noted, unexplained.

The problem of spherulitic crystallization, according to Keith and Padden, falls into two parts: firstly to account for the development of a radial fibrosity and secondly to explain the proliferation of fibrous units to fill space uniformly and maintain spherical symmetry. The latter phenomenon they describe as small-angle non-crystallographic branching, which is a major distinction from dendritic crystallization where parent and daughter fibrils share a common lattice orientation and have a precise mutual orientation. In spherulites there is no such precision: they are polycrystalline entities with the relative angles between branches generally increasing in frequency of occurrence as their magnitude decreases. As will be seen, this behaviour is not difficult to rationalize, but first the development of fibrosity is considered.

The nature of the problem is indicated by the constancy of the isothermal growth rate. Crystal growth is usually governed by diffusion, either of heat or matter, yet the solution of the diffusion equation for an expanding sphere is known to give a surface area increasing linearly with time, t, i.e. the radius as $t^{\frac{1}{2}}$. If diffusion is the dominant process, therefore, the implication of the constant growth rate is that local conditions are not changing with time.† Keith and Padden, therefore, concentrated on the processes of local diffusion at the crystal/melt interface and noted, further, that in certain systems the melt left between the growing fibrils of a spherulite never crystallized on subsequent cooling (Fig. 2.1c). They suspected, correctly, that this implied that the condition of the melt varied from point to point and formulated a theory based on the diffusion of impurities away from growing crystal surfaces.

Work of Rutter and Chalmers on the growth of metals from impure melts had shown that, under certain conditions, the propagation of a plane interface became unstable and columnar projections developed. Although a polymer sample might appear to be a chemically pure system, Keith and Padden recognized that it is in fact polydisperse; it may also contain atactic species, branched or entangled molecules. All of these elements will lead to variations in growth rate among different species. The less easily crystallized will tend to build up in numbers at a growing interface, i.e. act as effective impurities, and then diffuse away to regions of lesser concentration. In other words, it was possible to extend the work of Rutter and Chalmers to polymer systems.

Keith and Padden's proposal is essentially as follows. Consider a supercooled polymer melt within which crystallization is proceeding at finite rate. Although there must be local temperature gradients to allow the heat of crystallization to be dissipated, the uniform linear growth rate shows that these must be small and the system may be regarded as effectively isothermal. Let there be an equilibrium concentration c_∞ of effective impurities at points remote from growing crystals and consider the distribution of such impurities adjacent to a hypothetical planar crystal surface growing into the melt. This must take the form shown

---

† One might consider that variation of nucleation rates for different directions lies behind fibrillar development, but there is little to support this contention from work on single crystals.

schematically in Fig 2.9*a*. The distance within which this concentration is enhanced is determined by the length $\delta = D/G$ where G is the rate of advance of the surface and D the diffusion coefficient for 'impurities'. This is because the solution to the diffusion equation in one dimension, x, contains the term $\exp(-x/\delta)$; similar terms, probably with additional geometrical factors, will arise in more complicated situations. δ is, therefore, an important characteristic length expressing the condition of the system during growth.

The effect of this build up of impurities on further crystallization, depicted in Fig. 2.9*b*, is to depress the equilibrium melting temperature T_L (liquidus temperature) for melt with increased concentration of impurities below that it would otherwise have. The effective supercooling $\Delta T = T_L - T$ is correspondingly decreased and the thermodynamic drive for crystallization to proceed reduced. If, however, a protuberance developed on the interface it would find itself growing among fewer impurities, i.e. at greater supercooling, though at practically the same temperature, than the initial interface. It would, therefore, be stable and one sees that a planar interface growing under these conditions is inherently prone to break up into an array of protrusions.

So far this development is hardly different in polymers than for metals. What does distinguish the polymeric case is that, whereas in metallic systems the protuberances at interfaces remain small, in polymers they can be seen in certain cases (polypropylene, polystyrene, etc.) to develop into long units, which for convenience we shall call fibrils, without at this stage implying anything further about their fine structure than that they are much longer than their cross-section. This difference arises because only in polymers is the segregated material so reluctant to crystallize that substantial time and/or further cooling is often required for the interstitial material to solidify, if indeed it does crystallize at all. Fibres result, therefore, because the poorly crystallizable – or in the case of atactic impurities uncrystallizable – components segregate between them and delay or stop solidification of the intervening melt. The 'diameter' of these 'fibres' is also likely to be $\sim \delta$ because this is the scale of the diffusion field at each growing tip, which sets an approximate dimension to the adjacent units.

A quantitative test of the theory is now difficult. Values of δ are not too precise, because of uncertainties in D – G is known accurately – but the relation of δ to the scale of the radial units must be imprecise until such time as appropriate boundary conditions can be set to the diffusion equation. Nevertheless, the trend of behaviour is clear; G is known from data such as Fig. 2.4 and D increases with temperature so that the relative changes in δ can be assessed and compared with observation.

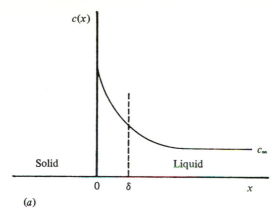

(a)

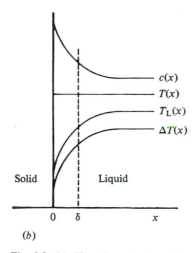

(b)

Fig. 2.9. (a) The schematic dependence of impurity concentration c with distance ahead of a growing crystal surface. (b) Corresponding variations in ambient temperature T, liquidus temperature T_L and supercooling $\Delta T = (T_L - T)$ produced by the varying concentration c; arbitrary units. (From Keith & Padden, 1963.)

For example, by choosing equal values of G at two suitable temperatures on either side of the growth rate curve (Fig. 2.5), the resulting variation in δ must be entirely that due to D and will increase for the higher temperature as is indeed found. Moreover, by keeping D more or less constant by mixing isotactic and atactic species of similar molecular mass, G is depressed for higher concentrations of additive (mainly owing to the decrease in supercooling) and this is found to increase δ and the scale of texture. Thus the spherulite of Fig. 2.1*c* was crystallized with 90% atactic additive and the texture is coarse (thick fibrils) and open whereas growth at the same temperature with less or zero additive produced appropriately thinner fibrils in more compact textures. The evidence of subtle textural and kinetic changes, explicable in terms of changes in δ and related growth conditions, which Keith & Padden (1963, 1964*a,b*) have presented, makes their case particularly convincing, but is not easily summarized and the reader is strongly urged to read their three main papers on the subject. It has also been shown directly, for example, by radioactive labelling of additives, that there is segregation between fibrillar units and between spherulites. This is confirmed by electron microscopy (see Chapter 4). The scenario of Keith and Padden has, accordingly, been generally accepted. It has sometimes been remarked that there may be other contributory effects, but convincing evidence to support this thesis has still to be provided. What may be more relevant, judging by the new electron microscope results on polyethylene spherulites, is whether such a simplification of the diffusion conditions is always justifiable and applicable to the structures observed.

In addition to fibrillation there must be small-angle branching to give spherulites their spherical symmetry. The fact of the matter is, principally, that small-angle branching becomes more profuse the thinner the fibrils. For very thick fibres such as those of Fig. 2.1*b* there is none at all, but in this case subsequent growth on the tips at lower temperatures (with much smaller δ) has been shown to produce profuse branching. The correlation of finer fibres and more frequent branching has been rationalized by Keith and Padden, who point out that it is consistent with new fibrils initiating at misaligned surface nuclei near the tips of existing units. If a new fibril should start it will only survive if its 'diameter' is commensurate with δ: smaller ones will disappear and larger ones break up. Moreover, the chance of there being a suitable nucleus must decrease as the requisite size becomes larger, so that for smaller δ there should be more branches and a finer texture, in the manner found.

2.3   Interaction of spherulitic and lamellar morphologies

Although the spherulitic and lamellar morphologies of melt-crystallized polymers have different origins the lamellae have to be accommodated within spherulites and it becomes inevitable that one must discuss how this is achieved. For the most part, lamellae in melt-crystallized polymers are discussed in the following two chapters and kinetics in

Chapter 6, but it is useful, nevertheless, to indicate in general terms what our understanding of spherulitic growth leads us to expect.

2.3.1   Kinetics

Spherulitic growth proceeds in two essentially different ways, which must be, and indeed are, reflected in isothermal growth kinetics. The first stage, *primary crystallization*, consists of the outward growth of fibres until impingement, the second, *secondary crystallization*†, which may well overlap the first, is growth filling in the interstices. The relative sizes of these two effects can vary widely, depending on the constitution of the melt and growth history; superficial morphology may tend to conceal the true situation. It is, for example, perfectly possible to produce a sample apparently completely full of spherulites which in fact contains 90% of uncrystallizable molecules, so that the additives must be concealed within the gross morphology.

2.3.2   'Fibril' morphology

In certain cases, for example open structures of polypropylene and polystyrene, optical microscopy can reveal radial units in spherulites which appear to be cylindrical fibrils (Fig. 2.1c). But this is not intrinsic to the theory just advanced, which merely sets a scale to the radial unit. In the case of polyethylene, Keith and Padden predicted that it was the width of a ribbon-like fibril which would be determined by δ, but the thickness would be set by lamellar considerations. What actually occurs is seen in Fig. 2.10. Fig. 2.10a shows how ridged sheets of lamellae, with an average of six ridge facets per major sheet, spread through the melt. It is clear that it is the width of the ridge which is related to the radial units (streaking) seen optically. The completed morphology is seen in profile in Fig. 2.10b, showing how this mode of growth has led to a quasi-periodic structure. Moreover, it is quite clear that in this case there were not fibrillar crystals growing in a honeycomb of melt, but rather the reverse, i.e. a matrix of ridged sheets enclosing 'fibrils' of melt. This appears to be a consequence of the lamellar habit. In these and other polyethylene spherulites the general details are much as expected on the Keith–Padden theory. It is becoming apparent, however, that, at least in polyethylene, the width of lamellae and the separation of regions of impurity remain within an order of magnitude of 1 μm and do not vary to the same extent as δ (Fig. 4.37). The significance of this is still hard to judge. It may well be that, for polyethylene, the simplification of the diffusion problem has gone too far. In other systems, notably polypropylene and polystyrene, the theory works well at the optical level. Not

† This term is not always restricted to this precise context in the literature.

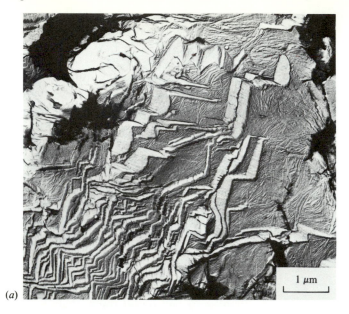

(a)

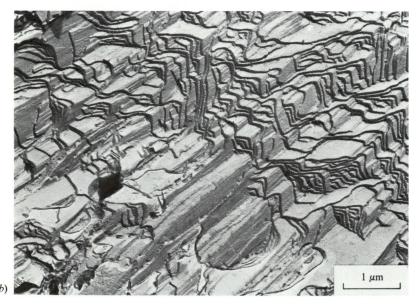

(b)

Fig. 2.10.(a) Ridged sheets grown at 130 °C parallel to the radius and at the edge of a polyethylene spherulite. The intervening polymer solidified on quenching. Replica of an etched surface. (From Bassett & Hodge, 1978b.) (b) Pseudofibrous appearance in a spherulite like the above but with all the lamellae having crystallized at 130 °C. Replica of an etched surface.

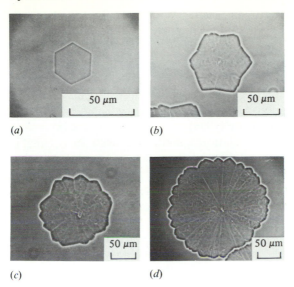

Fig. 2.11. The instability of polystyrene lamellae with increasing crystal size. (From Keith, 1964a.)

until electron microscopic investigations have been made on these as well is it likely to be possible to assess the validity of the theory in detail.

2.3.3   Limits of spherulitic growth

Spherulitic growth becomes operative, according to Keith and Padden, when the scale of the growing unit reaches $\sim \delta$, which for polymer melts is often $\sim 1\ \mu m$. For units of smaller size one would not expect a planar interface to break up. The point has been illustrated by Keith (1964a) in Fig. 2.11 for polystyrene growing in thin films close to the melting point where, because of large D, δ is correspondingly high. Here are four different crystals, clearly demonstrating how the ideal hexagonal shape degenerates once sizes of $\sim \delta$ are attained.

It is also implicit in the treatment of spherulitic growth that 'impurities' are localized by being unable to diffuse away from an interface at rates large compared to the growth rate. If this were not true, then the local diffusion fields responsible for spherulitic growth could not be stable and one would expect to observe a change away from linear growth rates to the parabolic laws pertinent to long-range diffusion. Such an effect has been observed by the deliberate introduction of highly mobile low molecular mass impurities into polymer melts. This will serve to re-emphasize a statement made at the beginning of this chapter, that spherulitic growth occurs under the influence of general crystal growth conditions when they happen to pertain to polymers.

2.4   Further reading

The optical properties of banded spherulites are dealt with comprehensively in papers published together in 1959, viz. Keith & Padden (1959*a,b*), Keller (1959*b*) and Price (1959).

A special recommendation is made to consult the original papers by Keith & Padden (1963, 1964*a,b*) on spherulitic growth. The first of these outlines the theoretical development for which detailed evidence is provided in the two subsequent articles. The scope, subtlety and elegance of the concepts presented are not readily summarized in a short presentation.

The technique of small-angle light scattering to study the size and organization of polymeric spherulites was introduced by Stein & Rhodes (1960) and has been used in many later papers.

3   Polymer lamellae

The next readily identifiable element of polymeric texture below the optical dimensions appropriate to spherulitic ordering is the lamella. This was discovered in 1957 for polyethylene crystallized from very dilute solution and is now recognized as a general textural component in most, if not all, crystalline synthetic polymers as well as certain biological macromolecular systems including DNA. Although it was then well known that crystalline polymers generally exhibited one or more low-angle maxima in X-ray diffraction patterns, the association of these with lamellar ordering was quite alien to prevailing opinion prior to 1957. What was especially astonishing was Keller's (1957) inference of a chainfolded molecular conformation within individual solution-grown lamellae. The repercussions of these twin discoveries, lamellae and chainfolding, have been felt throughout polymer science. They have, in particular, stimulated much research into the characterization of polymeric morphologies and their effect on properties.

From the beginning it has been appreciated that there must be a spectrum of ordering within lamellae. It is reasonable to expect, in general, that crystals formed from the melt, in which molecules are believed to be entangled and are likely to compete to crystallize, will be less ordered than those precipitated from dilute solution where molecules are well separated and may well add singly to a growing crystal surface. It has not been easy, however, to quantify such differences and understanding of melt-crystallized laminae has, for the most part, tended to proceed by analogy with those from dilute solution. In particular there is still no direct means of establishing the regularity of folding within lamellae; whether, for example, a molecule leaving a crystal returns at an adjacent site (*adjacent re-entry*) or not or whether the number of carbon atoms in a fold is small, say four or five, corresponding to ordered trans-gauche bond sequences as in cyclic paraffins (*tight folding*) or large (*loose folding*). These matters have become highly controversial. Nevertheless, the existence of lamellae is, by itself, sufficient to give much insight into polymer structure and properties, which can be affected most markedly. For example, the lamellae usually encountered are, say, ~20 nm thick and ~5 μm wide, although these dimensions can vary greatly, increasing by two orders of magnitude under certain conditions. When lamellae are microns thick, as in polyethylene crystallized anabarically at ~ 5 kbar (Chapter 7), then typical molecular lengths are commensurate with crystal thicknesses; the probability of tie molecules linking adjacent lamellae must be much reduced and samples become brittle. Moreover, it is appreciated that wide lamellar interfaces with a majority of van der Waals' bonding to their neighbours are mechanically soft and lower the Young's modulus of plastics by two or so orders of magnitude. If, however, the lamellae are disrupted to form narrow fragments mechanically linked in parallel with the soft interfaces, as probably happens on drawing to high draw ratios > 20, then the modulus can increase dramatically.

At present, the subject of polymer morphology has only just reached the stage of being able to characterize representative lamellar ordering in typical samples and has rarely been able to proceed beyond this to place the macromolecules in their textural context. A study of polymer microstructure, therefore, still tends to concentrate on the study of polymer lamellae. In this chapter those characteristics will be considered which tend to be common to lamellae in general while, in the next, more specific features and the mutual relationships between lamellae within semi-crystalline polymers will be treated, although this is not a division to which one can rigidly adhere.

3.1   Polymer single crystals

3.1.1   Preparation

Lamellae of most, if not all, crystallizable synthetic polymers can be grown simply using the method by which polyethylene laminae were first grown from solution. This was attempted, initially, as a means of producing an open texture more amenable to microscopic examination than the very compact melt-crystallized morphologies, which are notoriously difficult to examine. If the polymer is dissolved in hot solvent forming a very dilute solution, usually of concentration 0·01 to 0·1% (values at which polymer molecules are effectively separated from each other in the liquid, so minimizing complications in crystallization due to molecular entanglements), lamellar crystals usually separate out on cooling. For biological polymers it is likely to be necessary also to control the pH of the solution carefully to obtain lamellae. The crystals which result come in a great variety of habits and sizes within the general rule of crystal growth that the most regular crystals are those which form most slowly, i.e. at the highest crystallization temperatures. All such crystals are produced at appreciable supersaturations, often at 20–30 K below the equilibrium temperature. The free energy difference during

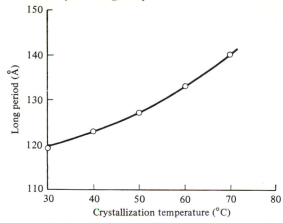

Fig. 3.1. The increase of long period with temperature of crystallization for poly(4–methylpentene–1) precipitated from 0.1% solution in xylene. (From Bassett, Dammont & Salovey, 1964.)

crystallization, i.e. between solution and crystal suspension, will increase as the temperature falls and, as in most solutions molecular transport is rapid, the crystallization rate increases monotonically with falling temperature. For controlled work it is best to crystallize isothermally; the temperature range available depends on the kinetics of nucleation and growth, as for spherulites, being smaller the more rapid these processes are. As an example the crystallization of polyethylene from 0·01% solution in xylene can be effected in the way described from about 90 °C, with crystallization times > 10 h, to about 70 °C when the timescale is one of seconds. This is a relatively large range because xylene is a good solvent for polyethylene: poorer solvents give smaller intervals. Despite the great variations in habit it is found that the thickness of the crystals formed at constant temperature is constant for the same solvent and virtually independent of concentration and molecular mass. For different solvents and the same polymer it is not temperature but the temperature interval below the equilibrium dissolution temperature which is the relevant variable. This straightforward dependence of lamellar thickness on crystallization temperature, independent of morphological complexities, is a feature which is amenable to theoretical prediction. The type of dependence with temperature (Fig. 3.1) is that given by nucleation. It corresponds to the precipitation of a liquid drop from its vapour where larger drops form nearer the equilibrium temperature because the additional free energy of the new surface must be offset by the reduction in free energy due to the condensation of the volume of liquid. The free energy available for compensation increases the lower the temperature is below the equilibrium temperature so that smaller and smaller drops then become stable. In consequence, nucleation theories of polymeric crystallization have been developed (Chapter 6) whose predictions are tested against data such as crystal thickness against temperature plots.

Such data is best obtained not from microscopic measurements of individual crystals but from small-angle X-ray diffraction of aggregated crystals. Suitable aggregates are prepared by filtering crystals from their mother liquor, first to a gel (at which point they may still be redispersed in concentrated proportions in fresh solvent, e.g. for microscopy) and then fully dried to a mat or cake. For small quantities it is easy to obtain homogeneous samples in this way, although too rapid filtration tends to orient crystals randomly instead of allowing them to settle slowly with their planes preferentially parallel to the filter. The manner of filtration can also significantly affect subsequent properties related to lamellar contact, e.g. response to nuclear radiations. For larger quantities, such as have been used to prepare macroscopic samples for tensile testing, it is necessary, after the gel state, to take careful precautions against the introduction of cracks by too rapid removal of residual solvent.

The preparation of polymer crystals by the method outlined above has certain disadvantages. Most serious is that the lowest crystallization temperature available is limited by the necessity of cooling the whole sample sufficiently quickly from some suitable value higher than temperatures where crystallization is known to occur. A lesser disadvantage is that the crystal population is often non-uniform in size and habit because of sporadic nucleation and molecular polydispersity. These can be overcome in various ways, though not without tending to produce countervailing drawbacks.

One valuable alternative is to prepare a concentrated solution, add this to, and disperse it within, a suitable volume of pure solvent already at the crystallization temperature (using mild stirring) so as to be effectively crystallizing in dilute solution. This tends to give a population of uniformly sized crystals all of which nucleated simultaneously at the moment of addition. The act of primary nucleation is, however, soon forgotten by the growing crystal whose morphology is governed by secondary nucleation at the growth front (as will shortly be demonstrated) so that crystals produced in this way still have morphologies appropriate to the crystallization temperature.

A second method which is especially valuable in exploring the variations of habit, because it restricts the molecular mass range used, is based on partly dissolving preformed crystals. For example, if a sample of polyethylene crystals is grown by the first method described, preferably slowly at 90 °C, then these are concentrated in suspension and plunged into a bath at, say, 96 °C only the edges of the crystals dissolve. The dissolved portion contains the shorter molecules from the original polymer and may be reprecipitated either as borders to the existing crystals or as new platelets. Examples of lamellae prepared in this way are seen in, for example, Figs 3.9 and 4.22. New platelets also tend to nucleate at zero time giving uniform sizes owing to the presence of microscopic nuclei in the suspension after dissolution. Experiments using samples fractionated in more conventional ways show sensitive changes in habit with molecular length. The method described reduces such complications by selecting a particular range of molecular masses; it proved particularly valuable in the early exploration of variations in habit with crystallization temperature in polyethylene.

A development of the above procedure has led to the *self-seeding technique* which is probably now the most widely used method of growing polymer lamellae because the entire population is composed of similar crystals. Continuation of the plunge procedure to higher temperatures (above about 97 °C for polyethylene) dissolves most of the crystals, leaving microscopic nuclei. These can sometimes be seen to have initiated growth at a subsequent, lower crystallization temperature (Fig. 3.2) giving, once again, a uniformly-sized population of crystals. Because of this instantaneous nucleation crystal growth is much more

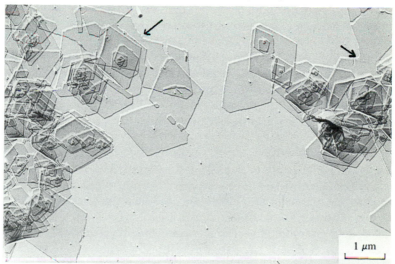

Fig. 3.2. Truncated lozenges of polyethylene grown by the self-seeding technique. Note the primary nuclei at lamellar centres. The arrows indicate where accidental notches have promoted growth.

rapid and it is not possible to crystallize at such low temperatures as in the first method (say below about 80 °C for polyethylene). Conversely, the highest temperatures at which crystallization can be effected are raised by some degrees. The particular value of this technique is, however, that it usually gives, at least for polyethylene, a population consisting entirely of monolayers whose size, moreover, can be controlled by varying the number of nuclei. Increasing the plunge temperature gives some control over this, higher temperatures giving fewer nuclei as more and more of the sample dissolves. For the most delicate variation, though, it is better to raise the temperature slowly by, say, 10 K min^{-1} in the range to about 105 °C. At the highest dissolution temperatures the number of nuclei is very sensitive to molecular mass and has, indeed, been used as a means of measuring the high tail of a mass distribution.

The nuclei certainly consist of long molecules, which light scattering suggests but, for reasons which are not understood, have a dumb-bell shape. The general conclusion is that lamellae are nucleated by the longest molecules in a sample and that there is then a progressive crystallization through the molecular range with the shortest molecules tending to crystallize last. Indeed, at the highest crystallization temperatures (> 85 °C for polyethylene in xylene) one commonly finds that the shortest molecules are unable to crystallize isothermally in reasonable times and only do so on cooling. Such samples show two X-ray long periods. Experiments in which crystals are filtered from their mother liquor at the crystallization temperature show that only the higher long period belongs to the precipitate; the lower value is confined to material of lower molecular mass which eventually precipitates from the filtrate liquor. This progression through the molecular mass range, generally from high to low, as growth proceeds is an important feature of polymeric crystallization not only from solution but from the melt. It has already been cited as being responsible for spherulitic crystallization and certainly exercises major influence on resulting properties.

3.1.2   Basic habits

Discussion of the details of crystal habits is deferred until Chapter 4 but certain basic features need to be appreciated at this stage. It is normally the case that the most dilute solutions produce the least complicated crystals. Concentrations of 0.01% will usually give monolayers, 0.1% will produce some multilayer development while 1% gives rather complicated structures already suggestive of melt-grown spherulites. Attempts have been made to approach the melt-crystallized morphologies by proceeding through more and more concentrated solutions but this, though a valuable exercise, may not achieve this aim fully because the solvent used is not continuous with the melt: xylene is always distinct from polyethylene. If, however, oligomers of the polymer are used as solvent then this otherwise inherent discontinuity can be avoided. Thus the crystallization of polyethylene from *n*-paraffinic solutions shows, even at 1% concentration, similarities to melt-grown behaviour which are still absent for concentrations as high as 30% in xylene. Specifically, the radial direction in polyethylene spherulites is *b* and outward growth along *b* is a feature of polyethylene precipitated from *n*-paraffins whereas growth tends to be outward along *a* from solution in xylene even to the highest concentrations which have been achieved.

Solution-grown lamellae are often called polymer single crystals. Certainly monolayers are single crystals in the sense of being separate entities; they are not, however, crystallographic single crystals, but rather multiple twins as will shortly be discussed. The number of habits developed is legion, even from a single solvent, and major changes can occur between different solvent systems. All the habits, nevertheless, tend, for a good solvent, to be variants of a simple shape based on the crystallography of intrachain packing. In polyethylene, which we shall need to discuss in some detail, the basic habit is the lozenge, a diamond shape outlined by four {110} facets, akin to that found for the ortho-rhombic *n*-paraffins (Fig. 4.7). Higher crystallization temperatures (> 80 °C in xylene) tend to truncate this lozenge by the introduction of two {100} faces (Figs 3.2 and 3.10*a*). Lower crystallization temperatures

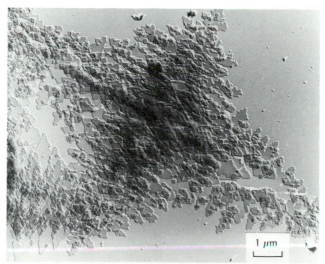

1 μm

Fig. 3.3. A polyethylene crystal which has become dendritic through the enhanced growth of corners and the proliferation of spiral terraces.

give a tendency to dendritic growth (Fig. 3.3) as the more rapid crystallization becomes controlled by diffusion of molecules to the growth front.

These changes can be seen most easily in the optical microscope, using phase-contrast or interference-contrast optics, by drying a drop of crystals in suspension on a glass slide. Such specimens can be transferred to the transmission electron microscope by evaporating a shadowing metal such as 40:60 Au:Pd alloy obliquely and a carbon film of about 20 nm (200 Å) thickness vertically on the slide. The composite film, of carbon plus shadowed crystals, may then be floated off on water in a Petri dish and portions collected on electron microscope grids in the conventional way. Cleanliness is essential and to aid in this it may be preferable to replace the glass slide by a thin, but rigid, sheet of freshly cleaved mica. A carbon film which resists floating off can usually be released by the addition of one drop of hydrofluoric acid to the water in the Petri dish.

3.1.3   The inference of chainfolding

The electron microscope not only adds detail to the optical shapes, confirming, for example, that dendritic crystals are indeed variations upon the basic lozenge-like habit, but also allows electron diffraction and diffraction microscopy. These are two of the most powerful tools available to the materials scientist and have provided much of our most detailed knowledge of crystalline texture (defect structures) in metals and inorganic substances. Carbon polymers, however, suffer from the serious disadvantage of being unstable in the beam of the electron microscope, undergoing crosslinking and/or scission of their chains while under examination. In consequence, the 'crystals' of polyolefines seen in the electron microscope rapidly become pseudomorphs of their originals, consisting very largely of heavily cross-linked networks of carbon chains. On the other hand lamellae of polyoxymethylene, which undergoes virtually complete molecular scission under electron bombardment, eventually leave little more than a replica of their original shapes, the irradiated polymer having disintegrated and been evacuated by the microscope's pumping system (Fig. 3.4). In addition, when large crystalline aggregates of polymers are examined, and especially melt-crystallized polymers, the irradiation necessary for examination produces complex changes, including mass transport, in the specimens and the images observed are by no means always simply related to the starting material. By comparison with inorganic materials, therefore, the detail revealed by the electron microscope in polymer samples is often severely restricted.

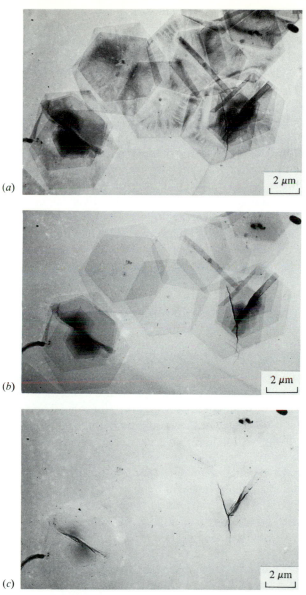

(a)

(b)

(c)

Fig. 3.4. Stages in the degradation of polyoxymethylene lamellae in the beam of an electron microscope. (From Bassett, 1964*b*.)

In the case of monolayers and simple aggregates, however, it has been possible to obtain a range of diffraction information by working rapidly at very low magnifications (~ 1000 times). By comparing the diffraction pattern of a lamella with its image, Fig. 3.5 (compensating, if necessary, for their mutual rotation within the microscope column), it is possible both to index a crystal's faces and to infer how the long chains pack within it. For polyethylene crystals, as their similarity to paraffins had suggested, the basic lozenge is edged by {110} faces, and truncated by {100} faces at higher temperatures. These indices refer to the subcell structure of Bunn (1939). The implication, pointed out first by Keller (1957), is that the molecules, which may typically be 0·5 μm long, lie across the thin (~ 10 nm (100 A)) dimension of the lozenge. It must follow that the chains fold back and forth repetitively as the molecule reaches each large or basal surface of the lamella.

This is the phenomenon of chainfolding. Since its initial discovery in 1957 it has been similarly inferred in virtually every crystalline synthetic polymer and even in more complex natural macromolecules such as cellulose acetate and DNA. By comparison with crystals of very short molecules (oligomers) which possess the same internal crystal structure – for example the orthorhombic *n*-paraffins in the case of polyethylene – it is evident that chainfolding is the new element conferred by increasing molecular length, which must ultimately have a major responsibility – together with the nature of intercrystalline connexions – for giving crystalline macromolecular solids their characteristically polymeric properties. Nevertheless, how the molecules fold is not determined by the evidence cited, only that they must bend back in some fashion. At first it was thought that, for example, folds in polyethylene would be relatively well-defined entities, as in crystals of the cyclic paraffins, with rather specific bond arrangements at the *fold surface* linking adjacent molecular segments (*fold stems*) in the crystal. Certainly, morphological detail revealed by electron microscopic study of monolayers and bilayers of various polymers shows compelling evidence for ordered arrangements of folding. It soon became clear, however, that there is equally strong evidence of disordered packing, even in the fold surfaces of monolayers, because the density of individual polyethylene crystals is about 0.98 g cm^{-3} compared to a crystalline density of 1.00 g cm^{-3}. There is, therefore, an essential ambivalence between ordered and disordered molecular arrangements in what are now thought to be surface regions (extending from the physical extremities back into the crystal interior) even in these particularly simple systems. The balance between these two factors shifts, as would be expected, towards increasing disorder with the increasing complications

(a)

(b)

Fig. 3.5.(a) Appearance of a multilayered lozenge-shaped polyethylene crystal. (b) Electron diffraction pattern of a polyethylene crystal similar to the above whose short diagonal was vertical. (From Keller, 1959a.)

of crystallization from more concentrated solutions and from the melt. Unfortunately, for polymers crystallized from the melt, the basic inference of chainfolding cannot be made in the way it was for solution-grown lamellae because it is not generally possible (except in special circumstances such as growth in thin films) to examine isolated lamellae; the necessity for folding molecules back into lamellae is, therefore, no longer self-evident. Nevertheless, such is the general similarity of behaviour between solution and melt-crystallized lamellae that the general consensus of opinion has been that differences in surface ordering between the two are of degree rather than kind. As has already been pointed out, there are still no direct means of characterizing fold structures; knowledge of them must always be inferred from a variety of indirect methods. At the present time the number of these techniques is increasing and for at least one new method, low-angle neutron scattering, some claim that results on melt-crystallized polymers are best interpreted in terms of minimal folding. In the course of the next few years it is likely that the facts of these matters will be resolved. In the meantime it is worth noting that there is no doubt that melt-crystallized polymers are lamellar in character and that in those cases where a straightforward test of whether molecular chainfolding is prevalent has been possible (crystallization in thin films, anabaric crystallization of polyethylene) the conclusions have been positive. In this volume the general view is taken that there is a continuous spectrum of ordering from solution to melt-grown polymer lamellae.

A progressive deterioration in traversing this spectrum can be detected, for example, from small-angle X-ray measurements of lamellar thickness. This specialized technique involves the measurement of diffracted rays deviated by less than 1 ° from the incident beam (i.e. for CuK radiation of spacings greater than about 10 nm). It is generally used to study the diffraction profile around the primary beam, i.e. the zero order of diffraction, which contains information on the size and shape of the diffracting units. Sedimented mats of solution-grown lamellae are unusual for polymeric samples in showing multiple diffraction peaks (frequently four for polyethylene) so that lamellar thicknesses l may be simply derived from the diffraction angle 2θ, for wavelength λ, using Bragg's law in the form

$$\lambda = l2\theta$$

Melt-crystallized lamellae, by comparison, usually show only one or two maxima, so that conversion of diffraction angles to the dimensions of diffracting units using Bragg's law is invalid, except as an approximate indication. There is no unique cause for the decreased number of maxima; it could be a consequence of disordered units (paracrystal-linity) or from a wide distribution in their size. The latter effect certainly exists for, although published data are still scanty it appears that, in contrast to solution-growth, crystallization from the melt gives lamellae whose thickness depends not only upon supercooling but also molecular mass, crystallization time and the environment of the growing platelet. For isothermal anabaric crystallization of polyethylene, even of good fractions with polydispersity < 1.2, the range of lamellar thicknesses covers at least one order of magnitude so that, at best, only weak maxima would be found in the small-angle diffraction of such samples. In this limited sense, therefore, ordering clearly declines for melt-crystal-lized samples and it is probable that this is also true of other aspects of ordering, as will shortly be considered.

3.1.4   Molecular orientation

The internal structure of polymer lamellae is conveniently approached by consideration of their internal molecular orientation. The diffraction patterns and associated lamellar shapes in Keller's classic 1957 paper on chainfolding suggested that molecules were normal to lamellae; it is not widely appreciated that for polyethylene this is not generally so. In point of fact polyethylene only rarely forms planar lamellae; monolayers grown from solution are often tent-like and, without special pre-cautions, they are flattened by surface tension when sedimented on a substrate prior to electron microscopy causing changes in their internal molecular orientation. If the observed diffraction pattern all comes from one specific area then the correct interpretation is that, in that area, the direction of molecules after deformation is normal to the platelet. The sharpness of the spots ensures that the diffracting regions are sufficiently extensive for the inference of chainfolding to remain intact. For initially tent-like crystals there is a further complication. These are multiple twins and the apparently single-crystalline diffraction patterns pro-duced are deceptive. They are, in fact, composites of spots derived from different twinned regions, vitiating the interpretation of a perpendicular molecular orientation. It turns out that molecules are not generally normal to polyethylene lamellae, but inclined at moderately acute angles to this direction. Chainfolding is still, however, correctly inferred from such a situation.

The above illustration shows how essential it is to associate diffraction patterns correctly with the diffracting regions and to avoid the complica-tions introduced by deformation of such flimsy lamellae. This is best done by dark-field electron microscopy. At the low magnifications necessary to study polymer crystals before their deterioration from beam damage becomes severe, this is obtained by displacing the objective aperture and recentring it on a selected diffraction spot (Fig. 3.6). Provided the aperture is of a size to admit only the one spot the image must be formed solely from electrons diffracted in and around the direction of the chosen spot and is, thus, bright on a dark field, whence the name of the method. The detail in the image is carried by the small-angle scattering of electrons around the diffraction spot, although this is not normally directly visible in the diffraction pattern. To define the chain-axis direction c one takes dark-field images through at least two non-parallel $hk0$ reflections and c, which is along the intersection of the two or more sets of reflecting planes, is then parallel to the electron beam in those regions which are present in all dark-field images.

A simple illustration of these methods is given by polyoxymethylene. Fig. 3.7a is a diffraction pattern of a lamella suggesting that the diffracting molecules in this hexagonal structure are parallel to the electron beam and, as the diffracting crystal is perpendicular to the beam, also normal to the lamella. This does not, however, rule out the possibility of a composite pattern resulting from multiple twins each containing inclined molecules, as occurs in polyethylene. Nor does a bright-field image (Fig. 3.7b) although it does reveal that essentially the whole

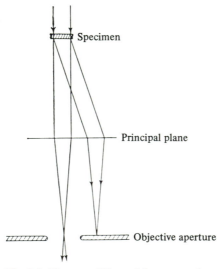

Fig. 3.6. Electrons diffracted from a specimen at a Bragg angle are usually stopped by an aperture, as shown, producing dark contrast on a bright field. If, on the other hand, the aperture is recentred around the corresponding diffraction spot, only these electrons can form an image which will thus be bright on a dark field.

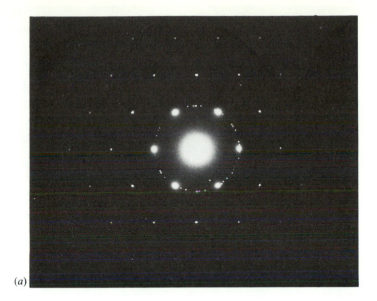

(a)

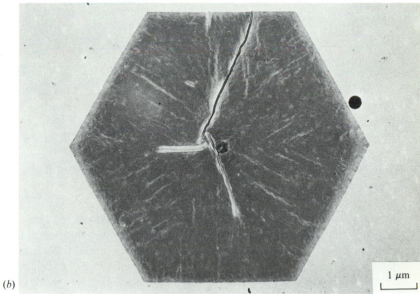

(b)

Fig. 3.7.(a) Electron diffraction pattern of a polyoxymethylene crystal. (b) A polyoxymethylene monolayer showing almost uniform contrast in bright field.

Fig. 3.7 contd over

(c)

Fig. 3.7.(c) A similar monolayer in 10.0 dark field corresponding to planes which run from eleven to five o'clock on the page. Notice the division of the lamella into six sectors and the absence of radial streaks for the two whose growth face is parallel to the diffracting plane. (From Bassett, Dammont & Salovey, 1964.)

crystal is diffracting. Dark-field images as in Fig. 3.7c (of a different crystal) do establish that the orientation of individual sets of 10.0 diffracting planes is maintained normal to the lamella throughout (with the exception of the radial streaks, whose significance will be discussed presently). In these (collapsed) polyoxymethylene crystals, therefore, molecules are essentially perpendicular to laminae; this is also true of their as-grown condition.

The same situation does not hold for polyethylene; lamellae of this polymer show different molecular orientations in different regions and, in individual regions, the orientation is rarely seen to be uniform. In a truncated lozenge crystallized isothermally, for example, there are six parts, called sectors, each extending from the crystal centre and bounded by the six individual growth faces. These sectors are readily distinguished by changing diffraction contrast in either bright-field or dark-field imaging (Fig. 3.8) or by small ridges along their internal boundaries (Fig. 3.9.)

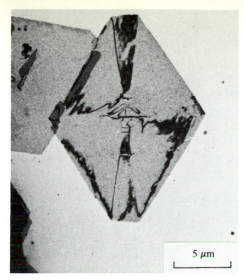

Fig. 3.8. Six sectors in a truncated polyethylene lozenge revealed in dark field. (From Bassett, Frank & Keller, 1959.)

Fig. 3.9. A truncated polyethylene lozenge with its six sectors delineated in bright field.

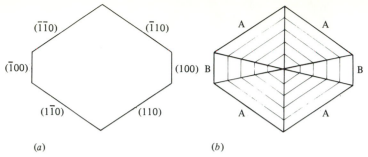

Fig. 3.10. (*a*) Face indices and (*b*) the pattern of folded ribbons in a truncated polyethylene lozenge.

3.2   Sectorization

The phenomenon of sectorization is most important for the understanding of polymer morphology, providing a fundamental argument based on symmetry for ordering of folded structures. At its most general the existence of distinct sectors implies that there is a structural legacy within each sector of the specific crystal face through which it grew. In most other materials this memory is lost but it can, for example, be seen in columnar chill crystals of metal ingots which nucleated at their outer surfaces and grew inwards in a similar way to the transcrystalline layers in polymer specimens described in the preceding chapter (Fig. 2.7*b*). For polymer lamellae, sectorization is almost certainly a consequence of the pattern of folding laid down at the growth faces.

Consider a truncated polyethylene lozenge as sketched in Fig. 3.10*a* growing from dilute solution. Molecules are most likely to add to the crystal by folding in strips parallel to the six growth faces according to the pattern of Fig. 3.10*b*. This would subdivide the lamella into six regions as shown according to the directions of the molecular sheets along the six faces. There is little doubt that this is so; the evidence is of three kinds. Firstly, when solution-grown polymer lamellae split, cracks parallel to the growth face of the sector are clean whereas others are often bridged by pulled threads in agreement with the molecular continuity sketched in Fig. 3.10*b*. Secondly, the chain packing is distorted according to the direction of the growth face, the sense of distortion changing systematically between different sectors. Thirdly, in an effect peculiar to polyethylene, molecules are inclined to the lamellar normals, but the tilt alters consistently from sector to sector. These points will be considered in turn following the introduction of suitable nomenclature.

3.2.1   Nomenclature for sectored crystals

A suitable nomenclature for simple crystals is that introduced by Bassett, Frank & Keller (1963a). In the case of a polyethylene lozenge there are four sectors, bounded respectively by the four growth faces (110), (1$\bar{1}$0), ($\bar{1}$10) and ($\bar{1}\bar{1}$0); we refer to these collectively as the {110} sectors and to that sector bounded by the (1$\bar{1}$0) face as the (1$\bar{1}$0) sector, and so on. One needs also to specify the *fold surface*, i.e. the plane of the basal surface in which the folds must lie. In polyethylene the fold surfaces are often close to {312} in which case the sectors would be of {(312)(110)} type. The six sectors of a truncated lozenge may be described similarly by {(312)(110); (*h0l*)(100)}. It is also useful to call the *fold plane* or *plane of folding* that plane in which the folded molecule lies; this is taken to be the growth face of the sector in the examples under discussion. The *fold length* is the interfold length, i.e. the interval measured along *c*, between successive folds in the same molecule while the intervening portion of molecular chain is known as a *fold stem*.

3.2.2   Subcell distortion

Very clear evidence that molecules have folded along the growth faces comes from measurements of subcell (i.e. chain packing) distortion. This is to be expected from general considerations of symmetry. Consider polyethylene: the ideal subcell structure of Bunn (1939) is orthorhombic so that in any one {110} sector the two {110} planes should be equivalent. This cannot be true, however, if there is folding along the {110} growth faces for, in the (110) sector (for example), (110) would be the fold plane, (1$\bar{1}$0) would not. The actual structure cannot then be truly orthorhombic but must be of lower symmetry, i.e. monoclinic or triclinic. The observed consequences of this are that the subcell structure is slightly sheared away from its orthorhombic dimensions, as sketched in Fig. 3.11, and a small difference is created between the interplanar spacings of (110) – the fold planes – and (1$\bar{1}$0) – the non-fold planes. For the

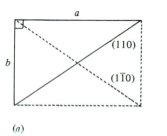
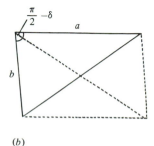

(*a*)   (*b*)

Fig. 3.11. (*a*) The (110) fold plane in the orthorhombic polyethylene subcell and (*b*) in the sheared cell appropriate to a (110) sector.

three polymers, polyethylene, polyoxymethylene and poly(4–methyl-pentene–1) the discrepancy is ~ 0.001 nm according to measurements using moiré techniques.

3.2.3 Moiré patterns

It is useful at this stage to digress partly and consider the phenomenon of moiré patterns. These take their name from the large-scale patterns observed when two silk sheets overlap. They are an example of a beat phenomenon, i.e. of the difference frequency between two interfering waves. When two polymer lamellae overlie suitably, such that their reciprocal lattices are nearly but not quite in register, then it is possible to observe a moiré pattern in the overlapping region arising from double diffraction of electrons by the successive crystals. Specifically, if the time-independent part of a plane diffracted wave of amplitude A is represented by $A \exp(2\pi i \, \mathbf{k} \cdot \mathbf{r} - \phi)$ where $\mathbf{k}$ is the relevant reciprocal lattice vector, $\mathbf{r}$ a direct vector representing position within the crystal and ϕ the phase at the origin, then successive diffraction by two such crystals allows a moiré wave of

$$A_1 \exp 2\pi i \, (\mathbf{k}_1 \cdot \mathbf{r} - \phi_1) + A_2 \exp 2\pi i \, (\mathbf{k}_2 \cdot \mathbf{r} - \phi_2)$$

Multiplication by the complex conjugate then gives the intensity variation in the image as

$$(A_1 - A_2)^2 + 4 \, A_1 A_2 \cos^2 \pi \, (\mathbf{k}_1 - \mathbf{k}_2) \cdot \mathbf{r} - \tfrac{1}{2}(\phi_1 - \phi_2)$$

This represents a series of fringes normal to $\mathbf{k}_1 - \mathbf{k}_2$ separated by an interval $M/|\mathbf{k}_1 - \mathbf{k}_2|$ (where M is the magnification of the image) which may be resolved provided $|\mathbf{k}_1 - \mathbf{k}_2|$ is sufficiently small. By considering double diffraction as occurring successively between all singly diffracted beams, and not just two as above, it is readily shown that, in place of a single diffraction spot from an individual crystal, double diffraction produces a superlattice, of the same symmetry as the direct lattice, based on the vector separation of the corresponding reciprocal lattice points of the two individual lattices (Fig. 3.12). An image constructed from beams associated with any one such cluster of superlattice points will thus be a pseudoimage† at low resolution of the original lattice (Fig. 3.13). Moreover, since this image is due to the difference between the two parents, any defect in one alone – such as a dislocation – will be transmitted to the image. This well-known general result has particular application to polymers because it allows the observation at low magnification of fine

---

† It is a pseudoimage because, although the superlattice has the symmetry of the parent lattices, the amplitudes and phases of the component beams do not correspond to those of the originals.

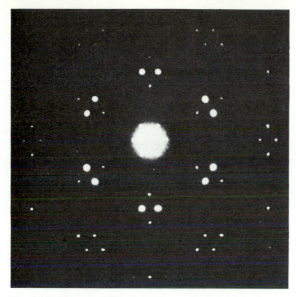

Fig. 3.12. Electron diffraction pattern from two polyoxymethylene lamellae which have a small relative rotation about their common chain axis. Identical patterns, but with a smaller rotation, produce moiré phenomena in the image at low magnification. (From Bassett, 1964*b*.)

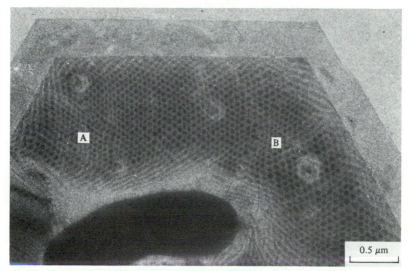

Fig. 3.13. Bright field moiré in polyoxymethylene revealing a pseudoimage of the hexagonal lattice and dislocations at *A* and *B*. (From Bassett, 1964*b*.)

detail, even to atomic dimensions, and thereby some easement of the restrictions imposed by beam damage upon direct imaging.

The moiré technique, used in dark field when it approximates to the two-beam situation just considered, has been employed not just to confirm the distortion of the subcell produced by chainfolding but also to measure its magnitude. It happens that an appreciable proportion of bilayer crystals tends to result when low molecular mass polymers are crystallized from dilute solution. In non-polymeric solids these two layers would be expected to fit one on the other exactly in register because of the identical pattern of atoms in the contacting surfaces. This is not, however, true of polymers, and the lattices of the two lamellae are generally offset by a slight rotation about their common normal; a sufficiently small rotation allows a moiré pattern to be resolved at low magnification. If it also happens that sector boundaries are not coincident in the two overlying lamellae, then lattice changes occurring across the sector boundary of one lamella can be measured against the (assumed constant) lattice dimensions of the other giving the results already quoted (Fig. 3.14).

The small discrepancy in dimensions between nominally equivalent planes of the subcell, depending on whether or not they are the fold plane of the sector concerned, means that the fold itself is not a random reversal of molecular direction. If it were, it would affect equivalent subcell planes equally. It follows that the folds have a preferred average conformation related to the fold plane. This would obviously be true were there adjacent re-entry and tight folding; how far this will persist as these conditions are relaxed is a matter for detailed calculation but this has not been attempted. It should be mentioned that a similar result is claimed from small-angle neutron scattering data of solution-crystallized lamellae which is interpreted as showing that individual molecules crystallize as sheets, presumably along the crystal surfaces.

Fig. 3.14. (*a*) Non-overlapping sector boundaries at five o'clock in a polyoxymethylene bilayer revealed in a 10.0 dark-field moiré. The chosen reflection corresponds to planes running from approximately one to seven o'clock on the page. (*b*) Interpretation of the above and calculation of the fractional difference Δd in interplane spacing and $\Delta \alpha$ in relative rotation. (From Bassett, 1964*b*.)

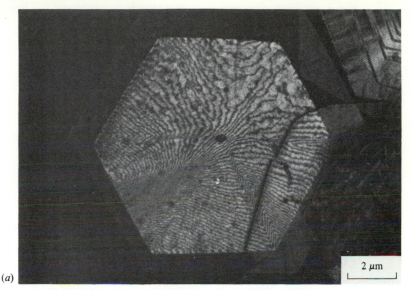

(a)

$d_1 < d_2$

A	B	C
$\alpha = 2.6 \times 10^{-3}$	$\alpha = 2.0 \times 10^{-3}$	$\alpha = 4.1 \times 10^{-3}$
$\frac{\Delta d}{d} = 0.5 \times 10^{-3}$	$\frac{\Delta d}{d} = 3.4 \times 10^{-3}$	$\frac{\Delta d}{d} = 0$

$d_1 > d_2$

$A \rightarrow B$	$B \rightarrow C$
$\Delta\alpha = 4.6 \times 10^{-3}$	$\Delta\alpha = 6.1 \times 10^{-3}$
$\frac{\Delta d}{d} = 2.9 \times 10^{-3}$	$\frac{\Delta d}{d} = 3.4 \times 10^{-3}$
$\frac{\Delta\alpha}{\sqrt{3}} = 2.7 \times 10^{-3}$	$\frac{\Delta\alpha}{\sqrt{3}} = 3.5 \times 10^{-3}$

(b)

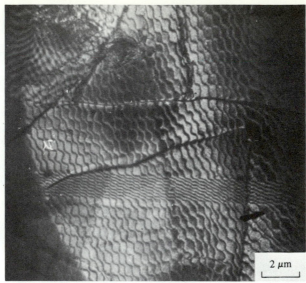

Fig. 3.15. Deformation moirés in a 110 dark-field micrograph of overlapping polyethylene lamellae. The central band is due to mismatched sector boundaries. (From Bassett, 1968*a*.)

For the lowest molecular masses, and particularly in polyethylene ($M < \sim 4000$), the $\cos^2$ profile of the two-beam moiré fringes sharpens and individual fringes instead of being straight lines become part of a polygonal network (Fig. 3.15). This is behaviour expected of two normal crystals in contact for it represents the resolution of the angular misplacement between the two layers into a twist boundary composed of families of screw dislocations. It should occur whenever the surfaces are sufficiently regular for improved packing between the dislocations to compensate for the cost of their additional line energies. Such moirés are called *dislocation moirés*; they are still formed by double diffraction but the fringes are now localized along the dislocation lines, i.e. in the interface, because this is where the phase changes between the two lattices occur.

The importance of deformation moirés is that the displacements between fringes are measures of the translation vectors of the fold surfaces, i.e. they give the distance between effectively equivalent matching positions in the surfaces. It turns out that for polyethylene, polyoxymethylene and poly(4–methylpentene–1) in all three cases the fold lattice has the same translations as the subcell. It is clear for polyethylene that this is inconsistent with random folding because this would make all fold stems equivalent, on a statistical basis, and $\frac{1}{2}\langle ab0 \rangle$ would be a transition of the fold lattice (instead of $\langle ab0 \rangle$ as observed) with the implication that alternate stems would be equivalent. Ordered folding involving adjacent re-entry could give this observed repetition; so also could chain-ends in the surface. As sectorization can tend to disappear under conditions when deformation moirés form in polyethylene it does seem that the presence of chain-ends in the surface is likely to play a significant part in the regular packing observed. Nevertheless, it does not appear justified to ascribe regular surface packing solely to chain-ends: (a) the deformation character of moiré fringes (narrow profile and kinked) disappears only gradually with increasing molecular mass and can be seen to change along the length of an individual fringe depending on the facility of local packing, (b) deformation character is clearly associated with moderate molecular lengths, especially in polyoxymethylene (Fig. 3.14) and (c) the conclusion of regularity in folding throughout the molecular mass range is indicated merely by the presence of sectorization as well as by other techniques.

3.2.4   Non-planar lamellae

Distortion of the subcell by fold packing makes polymer lamellae non-planar. Fig. 3.11 illustrates how the nominal right-angle subtended by a {110} sector of a polyethylene lozenge may, in fact, be $90° - \delta$. Fitting four such sectors together (without dislocations at the boundaries, in accordance with observation) would leave a gap of 4δ in a planar sheet. Alternatively, by forming a hollow-pyramidal sheet the four sectors can join smoothly producing a lamella which is a fourfold twin with each sector boundary a twin boundary.

Such crystals are only found for polyethylene of low molecular mass but the equivalent forms are typical for polyoxymethylene (a six-sided hollow pyramid) and isotactic poly(4–methylpentene–1) which gives a square-based hollow pyramid. Both are only slightly non-planar; the semi-angle of the pyramid for the latter polymer being only $\sim 88°$. Nevertheless, flattening on a substrate leaves a characteristic deformation, evident for polyoxymethylene in Fig. 3.7c, of radial streaks. This photograph shows that in the sectors whose growth faces are parallel to the diffracting planes there are no streaks, implying that in any sector the streaks are deformed regions which have rotated (sheared) about the normal to the growth face. This deformation does not disturb the orientation of planes parallel to the growth face but rotates all other {hk0} away from diffracting position. Such streaking may be taken as identifying subcell distortion and, with it, regularity in folding; it is present, for example, in cellulose triacetate as well as the cases mentioned above.

Polyethylene, however, yet again reveals a further facet of behaviour in that the non-planarity due to subcell distortion is usually masked by much greater non-planarity caused by molecular staggering. It is common for aliphatic compounds to form lamellae whose basal surface is not {001} but some other plane such as {201} which is found, e.g., in $n - C_{94} H_{190}$. Inclining the molecules to the basal surfaces in this way allows the end groups to occupy more surface area (Fig. 3.16). It may be achieved for such low-index planes by a progressive stagger of the molecules, for example for {201} by displacing every {200} plane by one c axis repeat distance. A similar situation occurs for polyethylene with

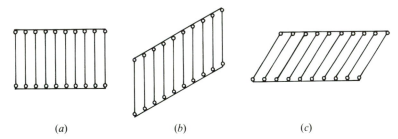

    (a)   (b)   (c)

Fig. 3.16.(a) Vertical packing of aliphatic molecules in a layer. (b) Oblique packing generated by uniform shear of part (a). (c) Part (b) rotated to give a horizontal lamella.

folds taking the place of end groups. It forms, for example, lozenges of {(312), (*hk*0)} type in which c is the axis of a non-flat-based hollow pyramid. Regarding c as the vertical direction this means that the contour lines would be the $\langle \bar{1}30 \rangle$ directions and not the traces of the growth faces. The significance of this feature is that not only must folded molecular ribbons in successive fold planes be displaced along c but there must also be a stagger of fold stems within a folded ribbon. This, however, would be expected for tight folding, on the following basis.

A useful model for discussing how a polyethylene molecule may fold back on itself and achieve adjacent re-entry into a lamella is the diamond lattice. This contains all possible paths based on tetrahedral linkages of covalent carbon–carbon bonds subject only to the constraint that the zigzag conformations of the two fold stems are parallel. Suitable bond sequences can be mapped out and then imagined subject to distortion to bring the two fold stems to the correct separation and relative orientation for the polyethylene lattice. Frank has suggested that the optimum diamond-lattice configuration to produce a fold linking adjacent stems in {110} in polyethylene is that of Fig. 3.17 (see Bassett, Frank & Keller (1963*a*)). This, and all alternative configurations when brought to the required orientation and dimensions, shows two kinds of asymmetry, namely left to right and front to back. If fold stems stagger to pack such fold shapes most economically, then left-to-right asymmetries would give a shear within a folded ribbon and front-to-back asymmetries would lead to displacement of adjacent ribbons, exactly in the way observed. It is entirely consistent with this argument that with lower molecular mass polyethylenes, for which the proportion of chain ends must increase, there should be no staggering and only non-planarity due to subcell distortion should be present.

The above considerations also lead to estimates of the additional free energy incorporated in a lamella by folding the molecule. The (major) contribution will be that of providing additional gauche bonds (see Fig. 1.6) – the fold in Fig. 3.17 contains four – plus the strain energy of

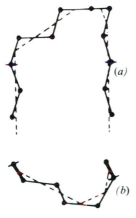

Fig. 3.17. Optimum path in the diamond lattice to simulate a {110} fold in polyethylene seen in elevation (*a*) and plan (*b*). Additional stem rotations and a small alteration in interstem separation are required to fit the polyethylene subcell. (From Bassett, Frank & Keller, 1963*a*.)

deformation. These suggest, for polyethylene, figures of ~ 60 mJ m^{-2} (erg cm^{-2}) similar to but a little less than those which are found in practice. Moreover, the free energy difference between different fold configurations is relatively small and is likely to mean that no unique fold will result from crystallization.† The same conclusion has resulted from detailed consideration of the observed fold surfaces: near {312} for crystallization above about 80 °C, near {314} for crystallization at 76 °C. (There are insufficient data to decide whether there is a progressive change of index or a discontinuous jump.)

---

† It may be possible to move towards the optimum by annealing.

The habit of polyethylene hollow pyramids indicates that folds in polyethylene tend to have preferred shapes. In view of the previous sentences, however, it has to be emphasized that this is an effect averaged over the whole surface. It is not likely to apply to all folds in a given surface, but only on balance.

The shapes of hollow-pyramidal and related polyethylene crystals were measured in Bristol in the early sixties by a combination of many techniques. The original papers should be consulted for details. The triangular pleat often found on monolayers (Fig. 3.8) suggests that lamellae observed in the electron microscope are collapsed structures. This was confirmed by sedimenting crystal suspensions on glycerine, instead of a solid surface. Carbon may be evaporated on this at a pressure of $\sim 10^{-3}$ mbar and, when the glycerine has been dissolved in water, may be shadowed then examined in the electron microscope. The angle of the pleat itself is a measure of the original shape, on the assumption that there is no shear during collapse but only rotation of the facets to lie on the substrate (Fig. 3.18). In the collapsed condition

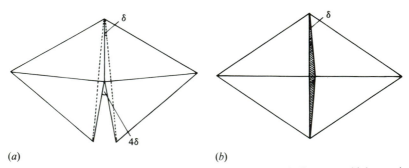

(a)   (b)

Fig. 3.18. Geometry of the pleat produced from a hollow-pyramidal crystal composed of sectors as in Fig. 3.11*b*.

the facets can be indexed by dark-field microscopy. The normal diffraction pattern is a composite in which each {110} sector only contributes to two of the four {110} diffraction spots (Fig. 3.19). The line between these two must, therefore, be near to the rotation axis of the sector. Rotation around this axis will bring c parallel to the beam and its amount determines the obliquity. The measured values were finally confirmed by direct viewing of uncollapsed crystals, floating in suspension, using optical dark-field microscopy.

3.2.5   Summary of morphological evidence

Microscopic examination has allowed the inference of chainfolding in solution-grown polymeric lamellae. It has shown by revealing distinct

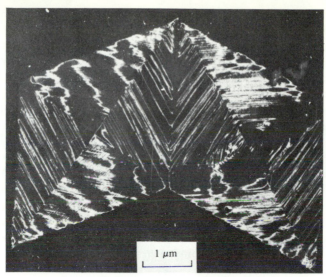

Fig. 3.19. 110 dark-field image, corresponding to planes which are vertical on the page, of two diffracting twinned polyethylene lozenges. Note that only two of each four sectors contribute substantially to the dark-field image. Contrast in the remaining two is limited to ridge crests (which will have deformed during sedimentation).

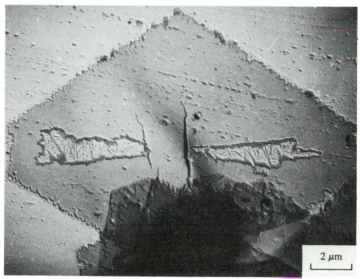

Fig. 3.20. A truncated polyethylene lozenge in which only the two {100} sectors have transformed following heating near the final melting point. (From Keller & Bassett, 1960.)

sectors (in which there is subcell distortion) that folds are not randomly placed but are likely to lie on average in the planes of the growth faces. Moreover, the non-planar habits of polyethylene hollow pyramids imply that, on balance, folds have specific shapes while moiré patterns provide evidence for adjacent re-entry. As a final confirmation of these conclusions there is the effect of thermal instability shown in Fig. 3.20. A truncated polyethylene lozenge contains six sectors; there are four {110} each in twin relation to the others and two {100} in twin relation to each other. The {110} and {100} sectors will have different structures because the folds will tend to lie along these different faces respectively. There will be different subcell distortions and different staggerings. All these factors will affect the thermal stabilities of the two types of sector differently. Fig. 3.20 shows that, in fact, the {100} sectors undergo melting, or some related transformation, earlier as crystals are heated into the melting region. There can be no doubt, therefore, of the reality of ordering of folds in solution-grown polymer lamellae.

3.3   The density of polymer lamellae

The picture of polymer lamellae presented by microscopy is apparently similar to that of lamellae of aliphatic compounds with the folds substituting for end groups. It is, however, easy to demonstrate that this needs major qualification: the density of the polyethylene subcell is 1.00 g cm^{-3} at room temperature whereas careful measurements give the density of solution-grown monolayers in the region of 0.98 g cm^{-3} depending on thickness. Evidently, there is a substantial element of disorder to be accounted for. Fischer & Schmidt (1962) showed that, under many conditions, the density ρ was related to that of the subcell ρ_c by

$$\rho = \rho_c - \frac{\text{constant}}{l}$$

where l is the lamellar thickness (Fig. 3.21). This has the form required for the density deficit to be located at the fold surfaces and it is generally agreed that this is so.

It is possible, for example, to swell polymer lamellae with a good solvent. The thickness increases (by ~ 0.8 nm for polyethylene lamellae) and surface mobility can be inferred from the movement during swelling of minute gold particles applied to the unswollen samples as a form of surface decoration. Moreover, the absolute intensities of the X-ray long

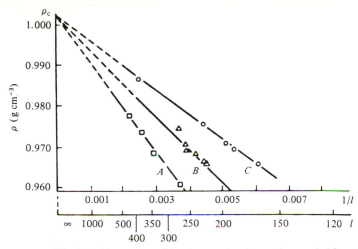

Fig. 3.21. Linear plots of density ρ against inverse long period for polyethylene single crystals following annealing for various times at A, 130 °C, B, 125 °C and C, 120 °C. (From Fischer & Schmidt, 1962.)

period can be related to the electron density deficit at the surfaces by the equations

$$\langle \Delta \eta \rangle^2 = (\eta_c - \eta_s)^2 \omega_c \omega_s = kI_o \int I(\theta) d\theta$$

applicable to a two-phase model of crystal interior and surface regions. Here η_c and η_s are the respective electron densities, ω_c and ω_s the corresponding volume fractions, k a constant and $I(\theta)$ the scattered intensity at θ from the incident beam of intensity I_o. The values of η_s obtained depend somewhat on the manner in which lamellae were aggregated but are generally close to values for the liquid phase extrapolated to ambient conditions. In short, there is strong evidence suggesting that the density of the surface regions, even of monolayers grown from very dilute solution, is close to that for an amorphous condition.

3.3.1   Organization of the surface layers

The surface regions of polymer lamellae evidently combine two very different characters: on one hand there are undoubted indications of order, on the other clear evidence of substantial disorder. Both must be accommodated in the same surface region. There is no inconsistency in assuming that this is the case. The evidence for regularity refers to average properties and cannot be claimed as showing that all folds are ordered. Moreover, the extent of the density deficit below the ideal crystallographic figure is too great, even for the highly crystalline linear polyethylene, to be accounted for on the basis of reasonable values for the densities of tight folds (which will be less than that of the subcell). In the case of branched polyethylene (where all but methyl branches appear to be excluded from lamellae) and other polymers of low crystallinity there is likely to be a considerable proportion of relatively disordered material at, or adjacent to, the fold surfaces. The difficulty is to be specific as to how this and the ordered fold pattern are arranged side by side. There is little evidence which bears directly on the point. Nevertheless, some indication of the likely answer is probably indicated by studies employing nitration/GPC.

Fuming nitric acid penetrates polyethylene, severing chains between lamellae, removing surface regions and digesting lamellae reducing their area in the *ab* plane, by a process conveniently described as ablation. The effects are rather subtle being selective, for example, between different populations in suitable circumstances. To a first approximation, however, the residual molecular lengths should measure lamellar thickness, taking chain inclination into account. Fig. 3.22 shows GPC traces of digested solution-grown polyethylene monolayers. The two peaks relate to molecular lengths corresponding to single and double traverses of a

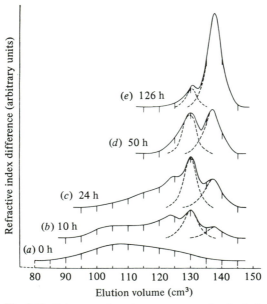

Fig. 3.22. Gel permeation chromatograms of polyethylene monolayers after degradation in fuming nitric acid at 60 °C for the times shown. (From Williams *et al.*, 1968.)

lamella; experimental values are in the ratio of 1:2 within 5% as would be expected for tight folding. However, the double peaks remain as more and more material is removed from the large surfaces, i.e. it appears that tight folding is distributed through a depth of a lamella associated with its surface. There have been other suggestions, among them the existence of an adsorbed layer of polymer: all, however, need to be treated with caution until the situation is clarified.

3.3.2   Melt-crystallized lamellae

When it has not proved possible to give an unambiguous description of the surfaces of the best-ordered lamellae grown from dilute solution, it is not surprising that the nature of melt-crystallized lamellae should be still more controversial. In part this is a consequence of few of the morphological, and especially microscopic, techniques used for solution-grown lamellae having been successfully applied to those crystallized from the melt. Nevertheless, it does appear that they, too, have a spectrum of ordering, including regular folding under optimum conditions, but increasing in irregularity for more rapid growth and longer molecules. The most regular habit is that of Fig. 2.10 in which the ridge sides are close to {201} orientation and, specifically, correspond one to one with the growth facets. This is sectorization of the kind found in non-planar solution-grown lamellae and is a similar pointer to ordered folding. There is no such evidence for most melt-crystallized polyethylenes but it should be noted that even in quenched morphologies there is a systematic textural pattern, contrary to the concept of random crystallization. At the present time, therefore, the evidence is in favour of a spectrum of ordering at the surfaces of polyethylene lamellae, grown either from solution or melt, which in both cases embraces regular folding, although the ordering is generally less for melt-grown crystals.

3.4   Melting point equation

The lamellar textures of crystalline polymers influence properties in a variety of ways, many of which are discussed in Chapters 8 and 9. One which needs to be dealt with now because of its fundamental importance to crystallization phenomena is the depression of melting point. The wide and variable melting range of a polymer is attributable in large measure to variations in lamellar thickness. The following derivation is, however, completely general for any crystal subdivided into lamellae.

The thermodynamic condition for solid and liquid phases to be in equilibrium at the melting point is that their specific Gibbs functions

$$g = u + pv - Ts$$

be equal.† The significance of the lower case symbols is that u, internal energy, v, volume, and s, entropy, refer to unit mass at pressure p and temperature T. Consider an infinitely thick solid of cross-sectional area A and density ρ which melts at $T_m{}^\circ$. Let $g = g_0$ at $T_m{}^\circ$ for the crystal and its melt. Now suppose the crystal to be divided into lamellae thickness l and cross-section A. Each lamella will have acquired $2A$ of new surface and $2A\sigma_e$ of surface free enthalpy where, for polymers, σ_e refers to the fold surface. Dividing by the mass $A\rho l$, one obtains $2\sigma/l\rho$ as the increase of g per lamella. This will depress the melting point to T_m. Recalling that

$$(\partial g/\partial T)_p = -s$$

and assuming that s may be regarded as constant near $T_m{}^\circ$, one has as the new condition for melting

$$g_0 + s_L(T_m{}^\circ - T_m) = g_0 + s_S(T_m{}^\circ - T_m) + \frac{2\sigma_e}{l\rho}$$

with subscripts L and S referring to liquid and solid respectively. Substitution for the heat of fusion per unit volume, Δh, where

$$\Delta h = \rho T_m{}^\circ(s_L - s_S)$$

and rearrangement gives

$$T_m = T_m{}^\circ\left(1 - \frac{2\sigma_e}{l\Delta h}\right) \qquad (3.1)$$

This equation is valid provided (i) the term $A\sigma_e$ is greater than equivalent terms relating to side surfaces, which have been neglected and (ii) that T_m is sufficiently close to $T_m{}^\circ$ for s_L and s_S to be regarded as constant. For polymers there are also additional terms related to possible molecular conformations in the solid but these are small and may usually be neglected. It should also be pointed out that l refers to the long period which is only equal to the fold length when molecules are normal to lamellae.

Equation (3.1) presents an apparently straightforward method of determining the important parameter σ_e from the slope of a graph of T_m against $1/l$. There are difficulties, however, in assuring on one hand that the temperature measured is actually the thermodynamic melting point and, on the other, that it corresponds to the recorded crystal thickness. For polymers these difficulties are particularly severe because of the metastability of chainfolded lamellae. On heating these transform into thicker crystals of higher melting point. If, therefore, one follows the

† The specific Gibbs function corresponds for a pure substance to the chemical potentials for multicomponent systems. It is now recommended that it be renamed the specific free enthalpy.

traditional practice of measuring melting points at as slow a heating rate as possible so as to approach most closely to thermodynamic equilibrium conditions, polymer lamellae will have changed considerably during the process. Conversely, if one heats very rapidly to minimize changes, one runs the risk of exceeding the true melting point, i.e. superheating crystals which become unable to nucleate and transform sufficiently quickly into the melt. This happens most readily for thick crystals and long molecules and has received considerable attention in anabaric polyethylenes and PTFE. In practice a compromise between the heating rates is needed to minimize the problems of transformation and superheating. Most often a temperature rise in the region of 10 K min^{-1} is recommended. In these circumstances T_m against $1/l$ plots (as in Fig. 3.23) give reproducible results and, gratifyingly, values of σ_e which compare favourably with determinations from other methods (Section 6.2). The intercept on the temperature axis is the parameter $T_m{}^\circ$ and gives values which tend to be a few K higher than the maximum melting points measured in practice.

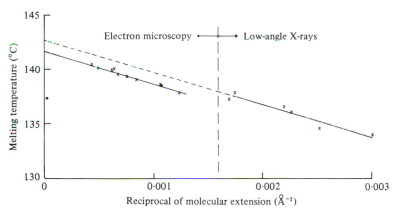

Fig. 3.23. Plot of melting point against reciprocal crystal thickness for oriented anabaric polyethylenes. (From Bassett & Carder, 1973.)

3.5   Further reading

A full description of the non-planar habits of polyethylene monolayers and discussion of their implications for polymeric growth may be found in Bassett, Frank & Keller (1963*a,b*).

4   Lamellar organization

The recognition of lamellae allows a convenient classification of the complexities of polymeric texture into (i) intralamellar features (microsectorization, twinning, etc.) and (ii) their arrangement in larger-scale structures such as spherulites. The latter topic, central to polymer morphology, is about as complete a textural description as is yet possible. For many polymers, especially when crystallized from the melt, only the broadest outlines of their lamellar arrangements have been derivable, e.g., by interpretation of the small-angle X-ray pattern. This is in terms of models whose validity must ultimately be supported by structures identified microscopically. Until very recently the relevant microscopy was likely to have involved replication of fracture surfaces, being virtually the only high-resolution technique available. Not only can this method obscure detail, however, but there have always been legitimate doubts about the representativeness of structures seen in fracture surfaces. Following the introduction of chlorosulphonation and permanganic etching these misgivings can be seen to have had some justification, at least for polyethylene. Certain populations, particularly of thinner lamellae, tend not to be revealed while those that do appear are viewed predominantly down the a axis because the preferential plane of fracture is {200}. The use of these two techniques heralds a new situation which studies, particularly using permanganic etching, have begun to exploit to disclose representative lamellar structures in melt-crystallized polyethylene. Detailed consideration is given to the techniques and their findings in Section 4.5.

The recognition of lamellae is only a limited objective, as platelets observed in different circumstances are likely to have different substructures. A wide variety of these is known for solution-grown crystals, and may reveal something of their history, but the internal structure of melt-grown lamellae is almost entirely unknown. Turning to the organization of multilamellar aggregates, one finds again that most of what is known derives from studies of growth from solution. There has been a number of investigations using the conceptually elegant and experimentally amenable means of approaching the state of organization produced from the melt by working with increasingly concentrated solutions. Although there is the important proviso that solvent and polymer are not continuous, this approach has literally allowed great

insight into the construction of crystalline polymeric aggregates. It tends
to reveal the interrelation of lamellae which may be considered a reasonable approximation to the description of the whole morphology. The
disordered regions present are usually associated with crystal surfaces,
certainly for the highly crystalline polyolefines which are the most
studied examples. The very name interlamellar frequently given to these
regions implies just such an association and is a tacit admission of the
paucity of information concerning their internal organization. For the
less crystalline polymers, the greater is the necessity to describe the
disordered regions in more detail to characterize the morphology sufficiently. However, current knowledge of moderately crystalline
polymers, such as polyethers, is sparse and one cannot yet go beyond
identification of lamellae and their mutual arrangements.

4.1   Lamellar microsectors

Polymer lamellae grown from solution are subdivided internally into
sectors. In the simplest cases these are outlined by boundaries drawn
between the corners of regular crystals and the common centre (Fig.
3.8). Such crystals are, however, only obtained under conditions of
relatively slow growth whereas those grown when a dilute solution is
allowed to cool naturally, or on quenching, tend to be more or less
dendritic (fir-tree shaped) (Fig. 3.3). Dendritic habits result when
growth becomes limited by one of its constituent transport processes. In
metals usually the dissipation of the heat of solidification is responsible
whereas in polymers movement of molecules to the crystals generally
controls the rate. Salient corners have a geometrical advantage, in this
situation, of being able to draw on molecules from a greater solid angle
than can neighbouring points on a growth face. Consequently, the corners tend to grow ahead of the remainder of the crystal, which becomes
attenuated, thereby further increasing the tendency to dendritic growth.

The change from nucleation-controlled to diffusion-controlled
(dendritic) growth is capable of infinite variation being susceptible to
the nature of the solvent, crystallization temperature, concentration,
molecular mass and other variables. Once again the changes are best
documented for linear polyethylene. These begin to occur, for 0.01%
solutions in xylene, below about 80°C, coincidentally about the crystallization temperature below which {100} faces no longer form. (All of these
values are subject to qualification because of the strong molecular mass
dependence.) As expected, samples with longer molecules begin to show
the effects of diffusion-controlled growth before those with shorter
molecules. For example, whereas the sequence of habits with decreasing

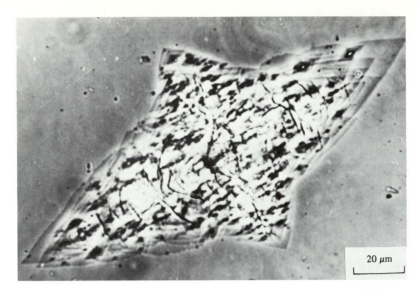

Fig. 4.1. A 'curved' crystal of polyethylene indicative of the onset of diffusion-controlled growth. Phase contrast. (From Bassett & Keller, 1962.)

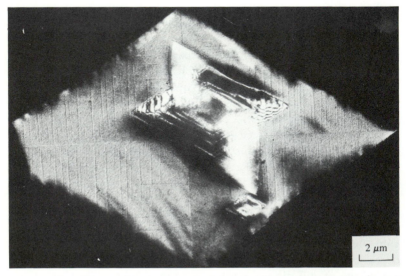

Fig. 4.2. Microsectors in a polyethylene lamella associated with re-entrant facets at the periphery; 110 dark field. (From Bassett, 1965.)

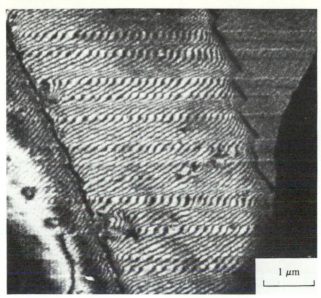

Fig. 4.3. Moiré fringes revealing lattice distortion across microsector boundaries in polyethylene; 110 dark field. (From Bassett, 1965.)

crystallization temperature is from truncated lozenges through true lozenges and 'curved' crystals to dendrites, true lozenges may not form if the molecular mass is sufficiently high.†

The first indication of the onset of dendritic growth is that the planar growth faces develop re-entrant facets. This happens for {100} at higher temperatures than for {110} faces but both can be seen in Fig. 1.3. Before long the crystal outline shows a characteristic curvature, identifiable by phase-contrast microscopy (Fig. 4.1). At higher resolution the associated small facets can be identified at the growth faces. These introduce microsectors into the crystal which in many instances (e.g. Fig. 4.2) can be traced back along the a or b axes towards, or even to, the major diagonals. Within these the folding pattern is changed exactly as between major sectors; this is shown by the repetitive alternation of moiré fringe directions across sequences of such boundaries (Fig. 4.3). This will, in turn, give rise to associated alternation of obliquity, and correspondingly crinkled lamellae (Fig. 4.4) for all except low mass samples. The length of the microsectors, originating on or near the major sector boundaries, indicates the stability of the microsectors, i.e.

† Most work of this kind was completed at a time when molecular masses were not easily quantified so that precise figures are rarely available.

0·1 µm

Fig. 4.4. Fine scale ridging revealing microsectors in polyethylene.

that the {110} growth faces have become unstable against the formation of re-entrant facets. With lower crystallization temperatures or longer molecules such effects increase, leading ultimately to classic dendrites with primary, secondary and tertiary branches (Fig. 3.3). When a lamella becomes dendritic during its development (Fig. 4.5) this is likely to illustrate changing conditions during growth as might be caused, for example, by a cooling sequence in which the increasing growth rate brought the onset of diffusion control part way through a crystal's formation.

It is also possible for sectors to be confined entirely within lamellae. One case is when, as in Fig. 1.3, an initially truncated lozenge continues to develop as a rhombus following a discontinuous drop in growth temperature. In suitable circumstances one can then detect transitional triangles, based on the original {100} faces, which link the six-sided and four-sided pyramids. Their straight sides trace the intersections of the facets concerned and are a further indication of the uniformity of the pyramidal slopes in a particular crystal. There tends to be a greater variation between the slopes of different hollow pyramids, consistent with the idea that the ordered folds in a given surface determine the magnitude of its obliquity, but subject to dilution of shear by less-ordered material.

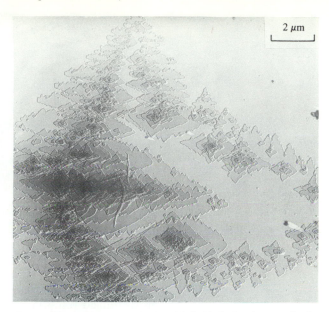

Fig. 4.5. The development of a dentritic habit during the growth of a polyethylene lozenge.

4.2   Non-planar monolayers

The hollow-pyramidal habits of polyethylene monolayers mentioned in Section 3.2.4 have a number of informative variants. Most widespread is the ridged lozenge, as modelled in Fig. 4.6. This form is related to hollow-pyramidal geometry by repetitive reversals of obliquity in the major sectors. The resulting ridges are generally continuous across the long (*a*), but not across the short (*b*), diagonal, Wider ridges are produced at higher crystallization temperatures. Growth between 75 and 80 °C has been found to produce intermediate forms including half-ridged, half-pyramidal lozenges.

Ridging a lozenge makes it stiff, as with corrugated paper, and modifies the mode of collapse of non-planar lozenges when sedimented prior to microscopy. Hollow pyramids usually tear or crease so that their excess material folds into a pleat (Fig. 3.9). If this is the only plastic deformation (except for bending at the inter-sector boundaries) then the original fold surfaces will lie undeformed on the substrate. This is the central assumption in reconstructing the undeformed shapes as {(312)(110)} hollow pyramids using orientations in collapsed lamellae established by dark-field electron microscopy. A similar mechanism

Fig. 4.6. Model of a ridged polyethylene lozenge. (From Bassett, Frank & Keller, 1963*a*.)

applied to ridged lozenges would lead to hinged rotation about, and eventual crumpling along, the $\langle 1\bar{3}0 \rangle$ ridge intersections. This does happen. In some instances, however, there is little or no sign of any crumpling and one concludes that collapse has then involved shear parallel to the common *c* axis which would project all the crystal smoothly on the plane of the substrate. In general, as indicated by the selection of sectors in dark-field imaging (Fig. 3.19) both rotation and shear deformation occur during collapse of ridged lozenges.

Phase-contrast optical microscopy is an excellent means of observing morphological details associated with collapse, especially if lamellae are watched before all mother liquor has evaporated. For example, Fig. 4.7*a* shows ridges in a true lozenge. The contrast is probably due to xylene temporarily trapped beneath the ridges, both of which ultimately disappear leaving only one prominent crumple (Fig. 4.7*b*). The truncated lozenge in this picture shows striations only within the {100} sectors. This is common and the occurrence of some shear which it suggests is probably a consequence of the collapse of the {100} sectors having to conform to the boundary conditions imposed by the larger {110} sectors.

Collapse can be avoided, to varying extents, by allowing crystals to sediment on a glycerine surface, as described in Section 3.2.4. This is particularly successful with ridged polyethylene lozenges which may be recovered apparently undamaged (Fig. 4.8). By comparison hollow-pyramidal lozenges are relatively flexible and only vestiges of their original non-planar shapes survive even this gentle sedimentation. In at

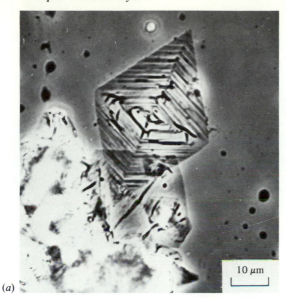

(a)

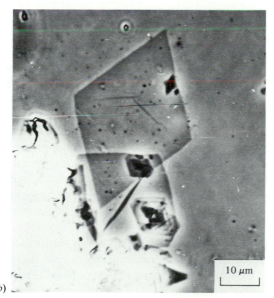

(b)

Fig. 4.7. Ridged and truncated lozenges of polyethylene (a) immediately after sedimentation from xylene suspension, and (b) two hours later. Phase contrast. (From Bassett, Frank & Keller, 1963b.)

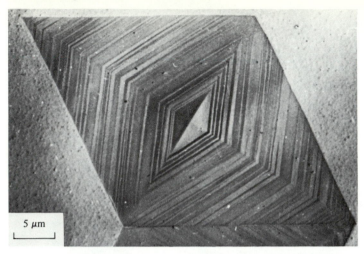

Fig. 4.8. Ridged polyethylene lozenge recovered uncollapsed after sedimentation on glycerine. (From Bassett, Frank & Keller, 1963*b*.)

least one instance, polyoxymethylene, it is also possible to observe the slight conicalness associated with sectors of a vertical molecular packing, using Nomarski interference contrast to observe lamellae on glycerine.

The existence of ridged polyethylene lozenges is intriguing. Each sloping ridge facet contains an ordered pattern of folding and is a microsector of $\{(110)(31l)\}$ type; alternate facets would be $(110)(31l)$ and $(110)(31\bar{l})$ for example. They differ from the microsectors described earlier by not having a different growth face for adjacent units: all have the same $\{110\}$ plane. It is likely that, instead of being grown into the crystal behind the associated microfacet, these microsectors have resulted from adjustment after growth, probably as folds interact and diffuse to make fold surfaces more regular and lower the free enthalpy of the lamella. This mechanism is suggested by dark-field electron microscopy of truncated polyethylene lozenges.

A hollow-pyramidal crystal, or *tent*, collapsed on a substrate without shear will then have different chain inclinations in each sector. These may be determined by appropriate tilting to bring *c* parallel to the beam, ascertained as described in Section 3.1.4. A suitable rotation should thus bring one sector into diffracting position at a time. This was observed and the numerical values indicated pyramids close to $\{(312)(110); (h0l)(100)\}$. At the same time other crystals were seen in which pairs of diametrically opposite sectors always came into diffracting position together showing that they had the same chain inclination. This is the *chair* form and is derivable from the tent by cutting the latter in half along the short diagonal, turning one half upside down and then constraining both halves to rejoin along the original cut. In the first experiments it was found that truncated polyethylene lozenges formed at 90 °C from 0.01% solution in xylene contained equal numbers of chair and tent forms.

Equal proportions suggest that lamellae begin to grow flat and that obliquity is adopted later, independently in each half. On the assumption that fold diffusion establishes the interaction leading to the adoption of oblique fold surfaces it, being a thermally activated process, will vary much less with temperature than growth rates which depend on supersaturation. One would then expect a spectrum of habits not unlike that previously described. For low supersaturation and very slow growth, fold interactions will develop before the addition of a further folded molecular ribbon. The tent form, which is presumably of lower free enthalpy, may then be selected, essentially at the nucleation stage, and its sense retained thereafter. All-tent populations have been found by Sadler (1968) for growth at 90 °C from 0.005% solution. With faster and faster growth one can envisage a progression of habits culminating in ridged lozenges, in a similar way to that observed. However, Sadler has also found all-chair populations (at 84 °C from 0·01% solution), a result which cannot be explained in this simple way without invoking further factors. The mechanism suggested is thus oversimplified but it is likely to reflect a major part of the answer and, indeed, is similar to mechanisms of fold smoothing required to explain annealing experiments (Section 5.3).

4.2.1   Relevance to melt crystallization

The complexities of microsectorization and its associated habits result from the sheet-like conformation of the polymer molecules crystallized from solution, i.e. folded ribbons parallel to the various growth faces. Such ribbons have also been inferred from small-angle neutron scattering of suitably deuterated samples of polyethylene crystals. A similar result has not been found for polyethylene quenched from the melt. For that, neutron scattering suggests that there are few or no sheets of molecules crystallized with adjacent re-entry.

There is very little direct morphological evidence bearing on this point but what there is is not in conflict with the above point of view. It has been almost impossible to examine melt-crystallized polymer lamellae by diffraction microscopy. Had this been the case also for solution-grown lamellae most of the evidence indicating regular folding would not have been available. However, a poor solvent already appears to give less regular folding from solution: bilayers of isotactic polystyrene so formed show moiré patterns but no sector boundaries.

A particularly suggestive series of experiments to illustrate melt-crystallized morphologies is that in which polyethylene has been crystallized from paraffin solutions, which may be regarded as from melt of very low molecular mass. As previously mentioned, the resulting habits share the radiating *b* axis characteristic of melt-grown polyethylene spherulites. Regular geometrical outlines appear in very dilute solutions (0.1%) with relatively large {100} faces, making *b* the direction of greatest extension (in contrast to *a* for lozenges precipitated from xylene). With increasing concentrations the outlines become curved but lamellae still are longest along *b* and associate in striking arrays, such as Fig. 4.9, which Keith (1964*b*) describes as floreate. Well-defined sectors are present when there are well-defined growth faces, and may be detected both by diffraction contrast and by ridges presumed to be a legacy of collapse. When faces become rounded this is accomplished by incorporating microfacets, particularly near the extremities, so that the laminae may become not merely lanceolate but acuminate, i.e. have a protruding point at their tips. There is much associated microsectorization in addition to a sharp change of slope across the spine of the lamellae corresponding to a change from (201) to (20$\bar{1}$) faces.

This habit is rather similar to that shown in Fig. 2.10 for polyethylene of $\sim 3 \times 10^4$ molecular mass crystallized at 130 °C *in vacuo*. It is a ridged sheet of about six alternating {201} facets, of which two adjacent faces would correspond closely to the previous example, growing outwards along *b*, the radial direction of spherulites. Each ridge facet appears to have its own distinct growth face, with symmetrical changes in these associated with a change of ridge slope. This has the appearance of a sectored habit and the same argument used in the discussion of hollow-pyramidal lamellae of polyethylene grown from solution will lead to the same conclusion viz. that there is a degree of ordering of the fold shapes and their placement. Ridged polyethylene crystals are grown very slowly (growth rates ~ 0.1 nm s^{-1}) from molecules of moderate length (i.e. few entanglements) and probably represent the optimum ordering resulting from melt crystallization. It would be expected that the most regular folding would occur in such circumstances. Lamellae crystallized more rapidly, or from longer molecules, are likely to be less regular – as their changes of habit confirm. Moreover, it has been observed experimentally that, unlike their solution-grown counterparts, melt-crystallized laminae only establish their full thickness a distance of approximately

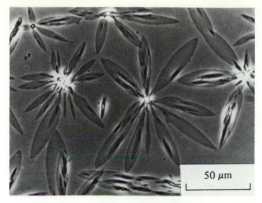

Fig. 4.9. 'Floreate' arrays of polyethylene lamellae ($\bar{M}_m = 4500$) grown from a 5% solution in n-$C_{32}H_{66}$ at 105 °C. The b axes are radial. (From Keith, 1964b.)

the same dimension, behind the growing edge. In consequence it is likely that initial fold placements become disrupted if not lost possibly leading to the disappearance of sectorization. Lack of sectorization, therefore, may well reflect this reorganization rather than a changed pattern of growth.

4.3   Twinning

The other major internal subdivision of lamellae besides microsectorization is twinning. Microsectorization itself does imply twinning of distorted subcells which, in the case of polyethylene, usually involves also changes of inclination of fold surfaces in oblique structures. Strictly this is twinning of the true unit cell, i.e. in which fold surfaces as well as interchain packing are considered. Twinning of a more conventional kind is that in which different components of a twinned lamella have the subcell in different crystallographic orientations and for which, at least to a first approximation, subcell distortion can be ignored.

A very considerable number of such twins is known from practically every system which has been examined, most of these being from solution. They are particularly prevalent in polymer of lower molecular mass, possibly because the relatively restricted number of primary nuclei gives greater scope for variations of growth to develop. Twinning is often instantly recognizable, and the twin law derivable, from the relative disposition of growth faces. Otherwise, and if possible, the assignments should be confirmed by electron diffraction. A selection of twins in polyethylene grown from solutions in xylene is shown in Fig. 4.10. The common twin planes in this system are {110} and, to a lesser extent,

(a)

(b)

Fig. 4.10. Twinned crystals of polyethylene: (*a*) sixfold star (phase contrast); (*b*) {310} twin seen in electron bright-field illumination. (Part (b) is from Wittman & Kovacs, 1970.)

{310}: this is similar to the behaviour of *n*-paraffins. Repetitive twinning across {110} planes can lead to a rosette type of twin with five or six arms. The angle between different {110} planes is 67° 30′ at room temperatures: fivefold repetition of this takes 337° 30′ and sixfold repetition 405°. Although these values are not very close to the ideal 360° required for

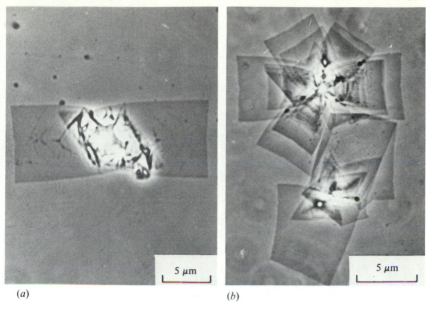

(a)   (b)

Fig. 4.11. (a) Lath-like {110} twin in polyethylene. Phase contrast. (b) Multiple twinning of lath-like twins in sixfold array, in a combination of features exhibited in Figs. 4.10a and 4.11a.

perfect matching in a monolayer, development in the third dimension allows suitable overlapping to produce the star shapes observed.

Most twins are relatively rare and may be regarded to some extent as curiosities which may, nevertheless, be informative. One such is the lath-like development of Fig. 4.11. This is based on {110} twinning with the twin boundary extending down the centre of the lath to a notch on the outer edge. Any notch tends to be a favoured site for nucleation. (Notice, for example, the enhanced growth at accidental re-entrants (arrowed) in Fig. 3.2.) In this case there has been more rapid growth along the twin boundary giving the strip-like habit growing out along ⟨110⟩ directions. Other lath-shaped lamellae of, for example, mono-clinic polypropylene or barium poly-*L*-glutamate, are not so easily explained. They have, indeed, become rather controversial because of an unresolved suggestion that chainfolding occurs parallel to their long axes out into the solution.

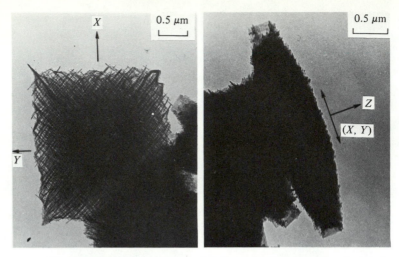

Fig. 4.12. Electronmicrographs of similar monoclinic isotactic polypropylene crystals grown from solution revealing their appearance in orthogonal directions. (From Khoury & Passaglia, 1976.)

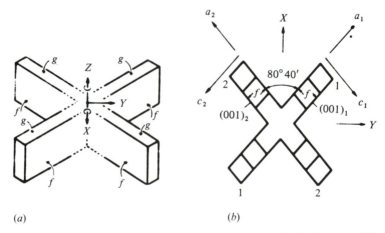

(a)   (b)

Fig. 4.13. Crystallography of objects as in Fig. 4.12. The fold surfaces {001} are marked *f* and the lateral growth surfaces {010} *g*. (From Khoury & Passaglia, 1976.)

Before leaving the subject of twinning, it is necessary to mention monoclinic polypropylene in which twinning is the norm rather than the exception. The morphology of this polymer is more complicated than most as Fig. 4.12 shows. These are among the simplest objects formed; even monolayers show incipient twin development at a very early stage. The crosshatched appearance is of lamellae viewed sideways on, but with two sets of layers in twin orientation; their crystallography is shown in Fig. 4.13. Unusually for polymer crystals the chain axis is not common between the two. The formation of this remarkable feature has not yet been explained, but it immediately leads to three-dimensional development and similar textures are representative of melt-grown monoclinic polypropylene spherulites.

4.4   Three-dimensional development

Recognition of the lamellar nature of crystalline polymers is sufficient to begin to understand the basis of many polymeric properties, but mechanical properties, especially, depend crucially on how the lamellae are connected. For these, and such fundamental textural problems of how two-dimensional lamellae fit within three-dimensional spherulites, one needs to know precisely how crystalline layers proliferate in the third dimension. Most of our knowledge comes from study of crystallization from increasingly concentrated solutions which yield increasingly complicated multilayer aggregates. In certain circumstances this has provided the only available systematic way of studying all perspectives of polymeric crystals. Moreover, the new information on representative lamellar organization in melt-crystallized polyolefines has tended to substantiate the relevance of such observations to melt-grown textures.

Nucleation of a new layer on top of an existing one, as envisaged in classical nucleation theory, does occur in polymers as the existence of the bilayers etc., used in moiré studies testifies. No doubt these are centred upon the primary nucleus of the initial layer which, according to theory, should be nearly twice as thick as the subsequently grown lamella (Section 6.1). This is not, however, a particularly common mode of growth – in Fig. 3.2 only monolayers occur even though the nucleating seeds protrude substantially from layer centres – except where low molecular mass polymer is involved. This is possibly because such molecules find it comparatively difficult to form primary nuclei; it is well known that longer molecules generally initiate crystallization more easily. Nevertheless, specially favourable locations such as exposed ledges are used as sites of preferential nucleation under suitable conditions. Fig. 4.14, for example, shows how small layers have started to grow along the second growth step, but not the first in a polyethylene lamella.† Both steps result from successive falls in growth temperature; that one is chosen as a site of subsequent nucleation but not the other, demonstrates clearly the delicacy of the selection process. More effective, in general, are the self-perpetuating ledges of spiral terraces produced by giant screw dislocations with Burgers vectors equal to the layer thickness (measured along the chain direction).

Growth pyramids are a familiar feature of lamellae grown from solution, and are especially prominent at low or moderate concentrations; they are also present to varying degrees in melt-grown specimens. Their origin is in accidents of growth and, in certain instances, these have been identified. One such is in the addition of a border to a {100} face of a polyethylene lamella at a temperature when only {110} faces form (Fig. 1.3). This leads to a row of spikes forming along what was a {100} face whose mutual displacement will initiate growth spirals, centred on the sector boundary, between adjacent spikes. The first suggestion was that this displacement was a bodily movement of a projecting spike but the same configuration can be reached, for nonplanar polyethylene lamellae, by a lateral translation of the sector boundary due to uneven growth: unless growth occurs symmetrically two sloping facets will not meet without overlapping. Such circumstances are very likely to prevail when growth is dendritic. Particularly revealing were the observations of Keith (1964*b*) on dendritic polyethy-

† Such crystals occur in mixed populations which include lamellae formed entirely at the lower temperature. The latter are usually wider than the border grown or added to existing lamellae. This is not necessarily an indication that it is easier to nucleate a new layer than add to an old one but more probably reflects the fact that the old lamellae have fallen to the bottom of the vessel under gravity and have to compete there for a more limited quantity of polymer than the new ones which will be randomly distributed.

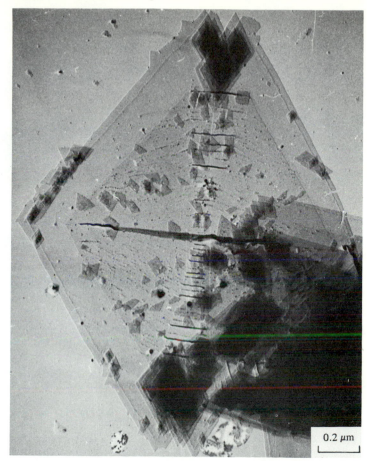

Fig. 4.14. Truncated polyethylene lozenge with two borders grown at successively lower temperatures. Note especially the overgrowths on the original fold surfaces only and preferred nucleation at the second step.

lene crystals exhibiting both salient and re-entrant corners between {110} and {100} microfacets. While, in principle, sideways movement of the sector boundaries during further growth at any such corner would initiate a spiral terrace as just described, in practice only re-entrant corners between a {100} and a {110} facet did so. The implication is that lateral movement of the boundary occurs more readily at asymmetric re-entrant corners. This is exactly what would be expected for diffusion-controlled growth because then the actual growth direction would tend to lie along the concentration gradient and bisect the re-entrant angle

Fig. 4.15. Screw dislocations generated at the edge of a microfacetted crystal of poly(4–methylpentene–1). (From Khoury & Passaglia, 1976.)

under the influence of competitive diffusion fields. Overlap at asymmetric, but not symmetric, re-entrants nor at salients (where the competitive influences are less powerful) should then result.

Although spiral terraces appear to be particularly prolific in the above circumstances it must not be supposed that their generation is restricted to non-planar lamellae. Evidently it is not, unless one includes in this term the slight conicalness due to sectorization, as Fig. 4.15 shows for poly(4–methylpentene–1). In this case microsectorization appears to be important, a point which is discussed more fully presently. In practice there are likely to be many ways in which adjacent parts of lamellae could become offset or overlapped and so generate growth pyramids. A further example is the 'picture-frame' development known in various polymers which is associated particularly with annealing in suspension and so discussed in Section 5.3.1.

Growth pyramids are the predominant way of multiplying the number of lamellae in polymer crystals grown from dilute solution (e.g. Fig. 1.3). They also occur in melt-grown lamellae,† though their relative influence has still to be established. In certain systems they are more numerous than others. The hexagonal modification of polypropylene, which is formed in thin films between ∼ 120 and 132 °C, has them in abundance (Fig. 4.16) whereas in bulk-grown polyethylene and poly(4–

† Except possibly in lamellae > ∼ 100 nm thick found in PTFE and anabarically crystallized polyethylene.

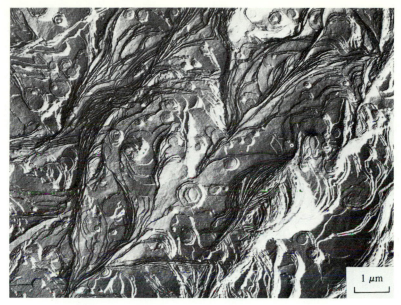

1 μm

Fig. 4.16 Morphology of hexagonal polypropylene crystallized as a thin film. Etched surface. (After Olley, Hodge & Bassett, 1979.)

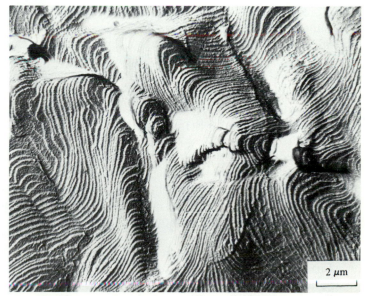

2 μm

Fig. 4.17. Internal morphology of melt-crystallized isotactic poly(4–methylpentene–1). Etched surface.

methylpentene–1) they are comparatively scarce. Indeed, the latter polymer exhibits linear stacks of rounded (as opposed to square) lamellae (Fig. 4.17) suggestive of nucleation along flow lines, even though the melt was not deliberately perturbed. Nucleation in strained regions where molecules are brought parallel for sufficiently long times is very important, not least commercially, but is discussed more appropriately in Chapter 7. It is pertinent to point out here, though, that it is not necessary to impose external stresses on a system to promote nucleation: internal stresses produced by crystallization itself can act in this way. The associated contraction will place the surrounding melt temporarily in tension and it has often been proposed that this could affect subsequent growth. Possibly the most suggestive supporting evidence has come from studies of rubber (cispolyisoprene) crystallized in thin films. This system has the advantages that growth can be stopped by quenching and the lamellae then stained with osmium tetroxide (which adds to the double bonds) permitting high-resolution electron microscopy. Often platelets can be seen close to, but separated by a few nm from their nearest neighbours. In the absence of direct contact it is particularly plausible that nucleation has been induced by strain.

In conventional laminar crystals, spiral terraces associated with a screw disclocation consist of contiguous platelets but this is by no means always the case for polymers. There are further influences responsible for shaping the three-dimensional habits. Examination, even of comparatively simple multilayer aggregates grown from solution, reveals a splaying apart of lamellar packets. Fig. 4.18 shows shapes (grown from ∼0.5% solutions of polyethylene in xylene at 76 °C) which strongly resemble those of immature open spherulites grown from the melt. These have been observed with phase-contrast optical microscopy while still suspended in liquid. Once collapsed the morphologies are much more confused, though electron microscopy still reveals lamellae and spiral terraces. Contrast in these phase-contrast images depends on the refractive index difference between liquid and crystal. For the combination of polyethylene and xylene this is quite small (the refractive index of both is ∼ 1.5), and considerable improvement may be effected by adding acetone (refractive index 1.36) to the mixture. An additional bonus is that this induces convection currents which can turn crystals over, revealing their different aspects successively. Systematic exploration of the appearance of crystals from all sides can be achieved by incorporating a liquid (such as aqueous PVA) which subsequently sets to a jelly and allows viewing with controlled tilting in a multiple-axis stage.

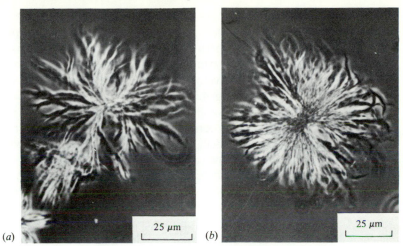

(a)   (b)

Fig. 4.18. Sheaf-like multilayer crystals of polyethylene observed suspended in xylene before collapse. Phase contrast. (From Bassett, Keller & Mitsuhashi, 1963.)

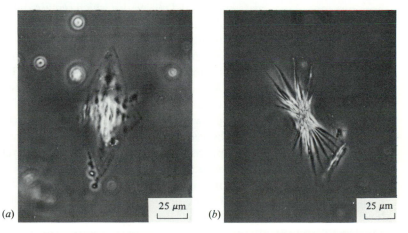

(a)   (b)

Fig. 4.19. The same splaying multilayer crystal of polyethylene seen flat (a) and edge-on (b) while suspended in liquid. Phase contrast. (From Mitsuhashi & Keller, 1961.)

A number of different polymers has been studied in one or other of these ways. including polyethylene, polypropylene, poly(4–methylpentene–1) and polychlorotrifluoroethylene. There appear to be two associated effects: splaying and curvature of lamellae with curvature increasingly evident the greater the development. Splaying without noticeable curvature is demonstrated in Fig. 4.19 for comparatively simple polyethylene multilayers; a section through a similar crystal is depicted in Fig. 4.20 showing the slender connection at the crystal centres. Even in polyethylene, however, the lamellae soon become curved, observed at the optical level, at relatively low concentrations, such as 1%.

Curvature has been studied in especial detail by Khoury, and in the case of poly(4–methylpentene–1), related to the distorting effect of chainfolding on the lattice and hence crystal geometry. The simplest square, four-sectored lamellae of this polymer are distorted (Section 3.2) by chainfolding and become approximately shallow cones with a semi-angle of about 89°. Khoury & Barnes (1972) have shown that, with increasing supercooling, re-entrants form along the growth faces which initiate new microsectored outgrowths (Fig. 4.15). Each of these should also have the same conicalness as the parent but, being located at the edges of lamellae, the total conicalness will be cumulative. This agrees with observation, and is able to explain the increasing curvature with lower growth temperatures through the increased frequency of microsector growth then created.

The origin of splaying itself is uncertain. An early and plausible suggestion is that this results from forces due to compression of cilia, i.e. those portions of otherwise crystallized molecules which remain outside lamellae and protrude from fold surfaces, e.g. chain-ends or long, loose loops. These must exist on statistical grounds and would be expected to be shorter when formed at lower temperatures. There is also substantial evidence that they are a significant feature of polymer crystals. They have been proposed as being responsible for nucleation of additional lamellae and, thereby, as the basis of a quantitative theory of the concentration dependence of growth rate from dilute solution (Section 6.1.3). It is also possible to observe related phenomena in the electron microscope. Specifically, it happens that a truncated lozenge of polyethylene whose edge has been removed by heating in xylene and then partly reprecipitated at a lower thickness and temperature, tends then to show small overgrowths, but only on the original fold surfaces and not on those of the new border (e.g. Fig. 4.14). Basal surfaces of polymer lamellae grown isothermally until the solution is exhausted of polymer molecules and then cooled to room temperature show no overgrowths. It appears, therefore, that cilia are not able to form new crystallites

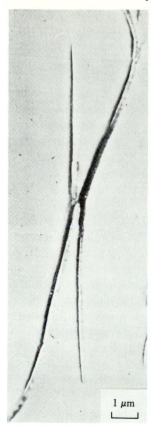

Fig. 4.20. Section through a splaying multilayer polyethylene crystal. (From Keller, 1967.)

unaided when they condense,† but can do so with the aid of additional polymer molecules from solution. This feature is proposed as the basis of the concentration dependence of growth rate, with cilia controlling the number of molecules accessible to a surface (Section 6.1.3).

Statistical calculations suggest that the lengths of primary cilia, i.e. chain-ends, are rather broadly distributed about a mean, of about one sixth of the molecular lengths, which is an increasing function of temperature. This applies to circumstances where chain-ends are excluded from lamellae which, while often the case, is not always true. When they are inside there are additional considerations: a chain-end is likely to cause a lattice defect, such as an interstitial or partly vacant stem, whose

† They probably contribute instead to the disordered regions associated with the fold surfaces.

energetic cost will disfavour short lengths. In agreement with this it has been found experimentally that, for polyethylene crystallized anabarically to lamellar thicknesses of ∼500 nm with molecular ends inside lamellae, the lengths of chain turned in from a surface are not less than ∼60 nm.

Whether cilia do, in fact, contribute to splaying is not known. It is noteworthy that splaying of lamellar packets in solution-grown crystals leads to branched morphologies very similar to those found in melt-grown systems. There the discussion has been in terms of the relative probabilities of nucleating daughter fibrils at small but finite angles to the parents. Neither point of view excludes the other, but no systematic study of branching and its contributory factors appears to have been made.

As splaying of solution-grown lamellae is visible optically, but individual lamellae are only, say, 12 nm thick, it follows that the mutually splaying entities must comprise groups or packets of lamellae. The obvious way to bring this about is to incorporate cilia from one lamella into its neighbours. This has long been anticipated and would give rise to tie-molecules, i.e. molecules bridging fold surfaces and being crystallized in neighbouring lamellae. The resulting incorporation of covalent bonds into what would otherwise be only van der Waals' bonding between fold surfaces has particular significance for mechanical properties.

There are at least two ways in which the existence of tie-molecules has been demonstrated experimentally. Historically, the earlier was a consequence of the discovery of the technique of *detachment replication*. In this, solution-grown multilayers are sedimented on a glass slide or, preferably, a freshly cleaved mica surface and coated with a 20 nm layer of evaporated carbon, with subsequent metal shadowing as desired. The carbon film can be floated off on water as described previously (Section 3.1.2) carrying the crystals with it and 2 mm squares collected on copper grids for the electron microscope. It turns out, however, that bringing such a film-covered grid into contact with the surface of hot solvent (xylene at temperatures ∼97 °C for polyethylene) removes all crystal layers except those in direct contact with the carbon. The top stratum, however, remains intact and still diffracts so that the technique permits one to detach just one layer for examination from what may be quite complex crystals (Fig. 4.21). It is rather analogous to investigating an onion by removing and examining just one coat. The physical basis of the effect is probably crosslinks formed between the highly energetic, hot carbon atoms arriving on the crystals and the polymer molecules: the technique works well for those polymers which form an appreciable proportion of crosslinks when subjected to ionizing radiations, for example, polyethylene and poly(4–methylpentene–1), but not for those such as polyoxymethylene which undergo virtually complete molecular scission. It is noteworthy, however, that, whereas dissolution of the inferior layers of solution-grown crystals is rapid and easy, occurring in seconds, this is not the case for melt-crystallized specimens which swell and distort. For these, dissolution, if it can be achieved without breaking the attached carbon film, is likely to take some hours. The probable reason is that in the melt-grown products there are many more tie-molecules connecting the top layer with its neighbours below. Even in solution-grown examples, however, it is possible to illustrate interlamellar connectedness directly by a more subtle use of detachment replication.

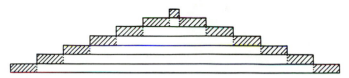

Fig. 4.21. Schematic of the surface layer removed in detachment replication. (From Bassett, 1961.)

When crystals are grown at two thicknesses following a fall in temperature, the thinner lamellae will dissolve before the thicker, according to equation (3.1). If, therefore, dissolution is carried out between the solution temperatures of the two, for example at 92 °C for lamellae 11 and 15 nm thick, the thicker layers will remain undissolved, in addition to the topmost layers of the detachment replica. In this way one can differentiate between growth above and below a lamella (Fig. 4.22). The inference of cilia may be made similarly. If an 0.01% solution of polyethylene is allowed just to begin to crystallize at 90 °C, by being held at this temperature for 12 h, and then to complete growth at 75 °C, the product is large, curved crystals. Detachment replication at 92 °C shows that these have often nucleated around a tiny truncated lozenge – a fact which would have been difficult to establish in any other way (Fig. 4.23). The seed crystals nevertheless remain in the specimens, evidently still linked to the outer surface although all intervening laminae have been removed. The only plausible explanation is that there are connecting cilia with lengths of tens of nm.

The second way of demonstrating interlamellar links is due to Keith, Padden & Vadimsky (1966) who cocrystallized mixtures of polyethylene and *n*-paraffins and then dissolved the paraffin away. This reveals threads of polyethylene bridging intercrystalline gaps of order 1 μm

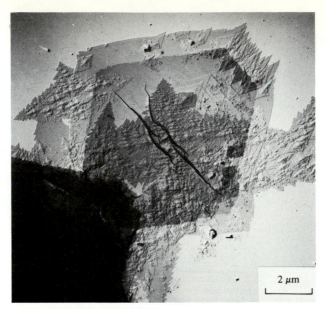

Fig. 4.22. Detachment replica at 92 °C of a truncated polyethylene lozenge 15 nm thick with a 12 nm border which overlies, and is overlain by, 12 nm thick dendrites. The contrast is due to the surface layer plus all other 15 nm thick layers. Note the growth pyramid growing through the truncated lozenge. (From Bassett, 1961.)

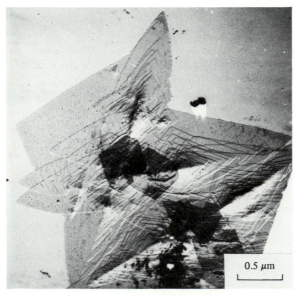

Fig. 4.23. Curved polyethylene crystals growing around twinned truncated lozenge nuclei. Detachment replica at 92 °C.

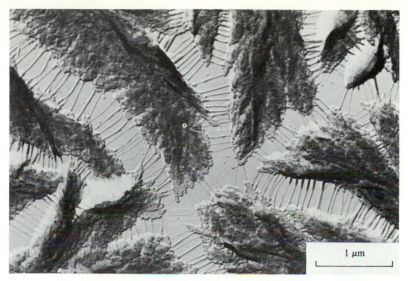

Fig. 4.24. Lateral intercrystalline links between islands of polyethylene grown from 50% solution in n-$C_{32}H_{66}$. (From Keith, Padden & Vadimsky, 1971.)

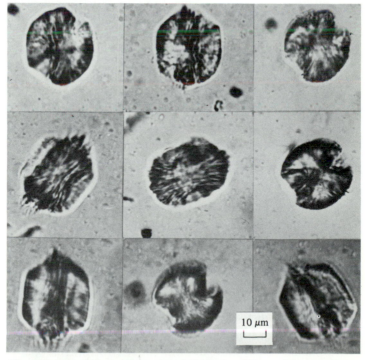

Fig. 4.25. Polyethylene axialite viewed from nine different directions. (From Bassett, Keller & Mitsuhashi, 1963.)

(Fig. 4.24). The links observed at ~ 10 nm wide comprise many molecules, all aligned along the threads. It is proposed that this results from molecules being pulled taut as crystallization proceeds at both ends but in different lamellae. If fresh polymer solution is allowed to crystallize on the threads it adds little platelets epitaxially transforming the appearance into what, for obvious reasons, have become known as *shishkebabs*. These can be formed in a variety of different ways, especially when crystallization occurs under strain, and will be discussed further in Chapter 7.

Interlamellar links can also be produced by deformation. It has already been noted that threads can be pulled across gaps in fractured polymer lamellae. Threads are frequently encountered in multilayer aggregrates – and for that matter in drawn fibres. It is by no means easy to decide, in a given example, whether these are a consequence of crystal growth (following Keith, Padden & Vadimsky), deformation of the sample prior to examination or inadvertent deformation during specimen preparation for microscopy (e.g. in stripping a replicating film off a sample's surface).

Returning now to the morphology of multilayer aggregates grown from solution, one observes that the overriding feature is the very different aspects presented in different directions. Fig. 4.25 shows the appearance of a single polyethylene *axialite*, so called because it consists, to a first approximation, of layers splayed about a common axis in the manner of the pages of a partly open book. The 'spine' direction is, in fact, not unique, but its angular variation is much less than the angles of splay about it. From most views, therefore, these objects resemble the sheaving shapes of immature spherulites. Nevertheless, the viewpoint most nearly normal to lamellae reveals a shape (allowing for the additional curvature) similar to that of solution-grown monolayers. Spherulites by comparison are believed to be collections of much narrower lamellae. The Keith and Padden theory relates this to the respective values of δ, which in solution is much greater than crystal dimensions so that morphological instability underlying a fibrous habit has not set in. In recent work the opportunity of investigating lamellar habits in melt-crystallized spherulites has been created. This forms the final section of this chapter.

4.5   Electron microscopy of melt-crystallized polyolefines

Two new techniques have transformed the possibilities of investigating melt-crystallized morphologies with the electron microscope. To recapitulate the difficulties involved, these are the fundamental limitations imposed by radiation damage, which not only destroys crystallinity but causes mass transport (which changes the appearance of the specimen) allied to intrinsically low contrast and difficult specimen preparation, especially the preparation of sufficiently thin ($\leqslant \sim 200$ nm) sections.

4.5.1.   Chlorosulphonation

The first technique is ostensibly a staining technique for polyethylene† to produce contrast by incorporating electron-dense atoms at lamellar surfaces. In fact it does this by converting the polyethylene into something else, with a melting point raised by ~ 15 K.‡ This is a chemically crosslinked polymer, with chlorine and sulphur atoms attached to lamellar surfaces, which has become easier to section and whose image contrast is relatively stable in the electron beam of the microscope. The type of image produced, illustrated in Fig. 4.26, has lamellae strikingly outlined in black (electron-dense) regions whenever their planes lie parallel to the beam. So far as has been ascertained this is a reasonable representation of the morphology of the orginal sample, although there are doubts about the measured lamellar thicknesses which appear to be too low. Less contrasting regions contain lamellae in other orientations (as appropriate tilting will confirm). With suitable specimens (whose thickness is of the same order as the lamellar thickness) the changes in image with tilt of the lamellar normal are particularly clear and can be identified by inspection. However, the tendency to observe lamellae most clearly when they are parallel to the beam makes this technique analogous to a dark-field method whereas permanganic etching, the second technique, by revealing lamellae in all orientations, would correspond to a bright-field image.

A typical treatment for chlorosulphonation would be to immerse a 2 mm cube of linear polyethylene in the acid at 60 °C for several hours to achieve homogeneity (detectable by the blackening accompanying the reaction, or by the melting endotherm). The times necessary decrease for samples of lower crystallinity (which the reagent penetrates more easily) but increase if reaction is carried out at lower temperatures, such as 20 °C, which may be necessary for drawn specimens which would otherwise tend to relax. The crystallinity of samples falls during chorosulphonation: for the highly crystalline polyethylenes crystallized anabarically at

---

† There are references in the literature stating that polypropylene also responds to chlorosulphonation, but it appears that the situation in that polymer is more complicated than in polyethylene and reversals of the expected contrast have been said to occur.

‡ This is probably a consequence of a reduced specific entropy of fusion, Δs, which is thermodynamically related to the melting point T_m and specific enthalpy of fusion Δh by $T_m = {}^{\Delta h}\!/_{\Delta s}$.

1 μm

Fig. 4.26. Structure in a polyethylene spherulite, similar to that of Fig. 2.10, after chlorosulphonation. Note the dominant and subsidiary ridged lamellae as well as intervening thin lamellae solidified on quenching. The view is down the spherulite radius.

5 kbar the decrease is to $\sim 50\%$ but, for example, with drawn samples, it may be impossible to retain crystallinity in homogeneously stained samples. The retention of some crystallinity is, of course, a prerequisite for the possibility of diffraction microscopy.

4.5.2   Permanganic etching

The second new technique is permanganic etching. Although etching is widely used in materials science to reveal salient morphological detail by selective removal of material from a surface, the many attempts that have been made to find suitable etchants for polymers, for example, based on solvents, their vapours, ions and activated gases, had met with only limited success. One of the best of these has been nitric acid which, following Palmer & Cobbold, (1964), is now widely used to separate polyethylene into lamellar fragments and, further, to remove their basal surfaces leaving only stems whose lengths are a measure of the thickness of the original crystals. As this severe degradation implies, however, nitric acid tends to be too strong a treatment although it has had some success, e.g., in indicating lamellar orientations in drawn material. The search for a suitable weaker acid has led to the use of potassium permanganate dissolved in sulphuric acid to concentrations of 7% or less which is capable of revealing lamellar detail at least in polyethylene, polypropylene and poly(4–methylpentene–1).

This reagent removes 1 to 2 μm selectively from polyethylene in typical treatments and, though it tends to attack interlamellar material preferentially, does not appear to penetrate far into the material, in contrast to nitric acid. This is an advantage for microscopy as the sample is not embrittled and can be handled relatively easily. One then has to examine the detail produced in the surface for which one has recourse either to two-stage replication or to high-resolution scanning electron microscopy.

4.5.3   Lamellae in polyethylene spherulites

An important feature of the way lamellae are organized within polyethylene spherulites is indicated in Fig. 2.10. Two types of lamellae may be distinguished. Initially, wide sheets advance into the melt (Fig. 2.10a), leaving narrow cavities to be filled in by later-forming platelets. The former are termed *dominant*, the latter *subsidiary* lamellae. The same distinction may be inferred from the lateral continuity of the ridges in fully crystallized regions (Fig. 2.10b) where the wider lamellae must have formed first. It is the dominant lamellae which will determine the growth envelope and ridged dominant lamellae form comparatively simple, though usually immature, spherulites which have been studied in detail.

Fig. 4.27 shows radiating structures, consisting of ridged sheets seen in various orientations around the radial *b* direction, in such a spherulite, with continuity often observable over ~ 30 μm, an appreciable fraction of the diameter which is typically ~ 100 μm. Closer inspection (Fig. 4.28) reveals local variations, to a few degrees, in the orientations of ridges. The implication is that some ridges will be more favourably orientated for subsequent crystallization than others in their neighbourhood. The better-placed will thus tend to gain in dominance and to establish the direction of growth to which late-forming lamellae must accommodate. Such a dynamic change of dominance can occur between different sheets or between adjacent {201} facets of the same sheet, it being likely that these are separated by tilt boundaries. In certain instances the variation of orientation can be caused by nucleation of daughter lamellae at angles to their parents, as occurs in Fig. 4.29. Here also is an example of how once subsidiary daughter lamellae can become dominant at a later stage of growth and provide a change in growth direction. The most suitable lamellae are thus selected for dominance by the growth processes.

There are three basic habits of dominant lamellae in polyethylene

Fig. 4.27. Radiating ridged sheets in a spherulite of polyethylene $\bar{M}_m = 2.1 \times 10^4$, $\bar{M}_m/\bar{M}_n = 1.7$ crystallized at 130.4 °C for 27 days. Etched surface. (From Bassett, Hodge & Olley, 1979.)

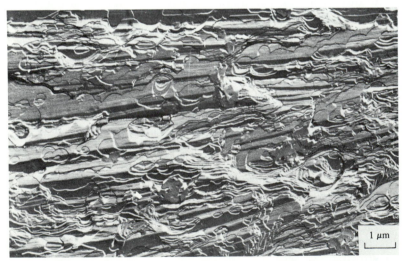

Fig. 4.28. Detail of Fig. 4.27 revealing the ridged structures. Etched surface. (From Bassett, Hodge & Olley, 1979.)

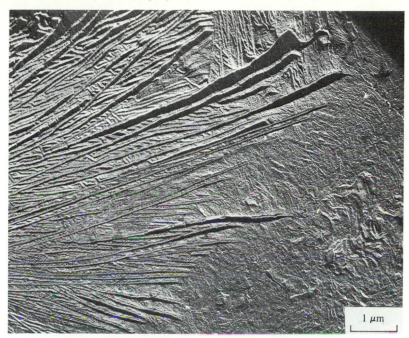

Fig. 4.29. Detail of a polyethylene spherulite. Note the facets at the tips of the growing ridged sheets (top) and the nucleation of daughter lamellae which eventually become dominant (lower centre). Etched surface.

crystallized *in vacuo*, differing in the profiles they present when viewed down *b*. In addition to {201} ridged sheets (fig. 4.30*a*) there are planar, or slightly curved {201} sheets and S-shaped sheets (Fig. 4.30*b*). It should be understood that the alignment is not exactly {201}, but is within a few degrees of this low-index plane, a similar situation to that observed for the {31*l*} facets of solution-grown hollow pyramids. The crystallography of all three shapes is sketched in Fig. 4.31 and shows that an S can be derived, to a first approximation, from a {201} sheet by suitable shear towards its edges. The near common *c* axis orientation means that there will be two locations on an S where molecules are normal to the lamellar surface. Elsewhere and in all other cases, molecules are inclined to lamellae in melt- (as in solution-) crystallized polyethylenes. This is undoubtedly a consequence of the small cross-sectional area per chain of the polyethylene subcell which would bring adjacent folds too close together in a vertical structure. The difference in shapes is possibly related to changes in the growth regime.

The sequence ridged, planar, S-shaped sheets occurs with increasing

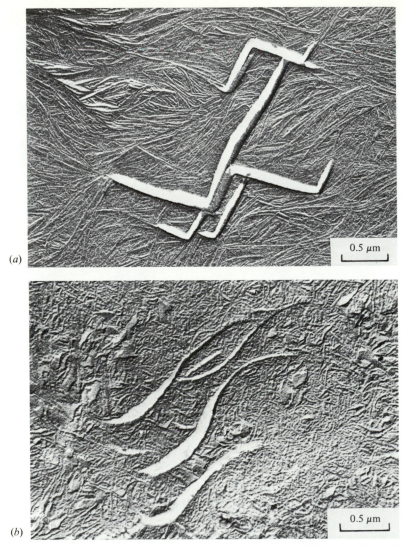

(a)

0.5 μm

(b)

0.5 μm

Fig. 4.30. Dominant lamellae in polyethylene revealed after quenching (a) towards the edge of a spherulite in the same etched surface as Fig. 4.27 (From Bassett, Hodge & Olley, 1979), and (b) after 1 h crystallization at 128.1 °C for a fraction with $\bar{M}_m = 4.4 \times 10^4$, $\bar{M}_m/\bar{M}_n = 2.0$. Etched surfaces.

molecular length (and thus growth rate) at 130 °C and again, but shifted to lesser lengths, at 128 °C and slightly lower temperatures. One may compare this with kinetic data (Section 6.2) which divide into two regimes for sharp polyethylene fractions. Dominant ridges fall within

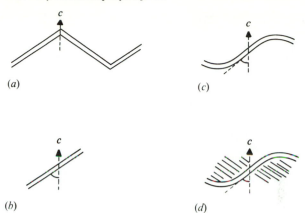

Fig. 4.31. Crystallography of dominant lamellae in polyethylene viewed down *b* and with the *c* axis vertical: (*a*) ridged, (*b*) planar and (*c*) S-shaped sheets. In (*d*) the pattern of subsidiary platelets associated with an S is also drawn. (From Bassett, Hodge & Olley, 1979.)

regime I which is interpreted as having well-separated growth centres along molecularly smooth growing surfaces. This is consistent with the morphology observed if each growth centre is associated with a single growth facet and thus a single ridge side. It would follow that facet widths of ~1 μm would be an upper limit for the separation of growth centres, in excellent agreement with predicted values. It is also likely that the change from ridged to planar sheets is linked to the disappearance of individual facets on the growth front. Such a change could well be expected for the onset of regime II kinetics within which, it is believed, there is profuse nucleation with rough growth fronts. The conditions giving dominant planar sheets are, however, rather broader than the abrupt change between kinetic regimes. It has still to be established whether this is simply because of the different polydispersities of the respective experiments or reflects deeper causes. Nevertheless, as dominant Ss are only encountered within regime II there does appear to be at least a coarse correlation between kinetics and lamellar morphology.

The complicated organization of dominant S-shaped sheets and associated {201} subsidiary platelets (Fig. 4.31*d*) prevails over most of the accessible growth conditions. Although no clear reason has yet emerged for the formation of Ss, which seem unique to polyethylene, one can recognize a link with banded spherulites. When Ss are first encountered at higher growth temperatures they have no overall orientation and there is no optical banding (Fig. 4.32). When there is banding then

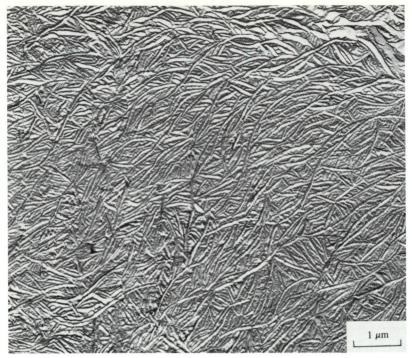

Fig. 4.32. A later stage of the morphology of Fig. 4.30 *b* after crystallizing completely at 128.1 °C. Note the somewhat disorganized arrangement of S-shaped sheets. Etched surface. (From Bassett, Hodge & Olley, 1979.)

neighbouring Ss have the same orientation so that they and their subsidiary platelets will possess a near common *c* axis orientation in the neigbourhood of a given point which the optical properties require. The organization looking down the radius of a banded spherulite is shown in Fig. 4.33 which should be compared with a diametral view of a similar spherulite in Fig. 4.34. The dominant Ss are apparent in both and are related to the sense of optical twist in such a way that, travelling outwards, an S would have to rotate so as to scoop up the melt. In fact there is little evidence of twisting of individual lamellae. The dominant Ss, and by inference their associated subsidiary platelets, appear to be essentially untwisted and to maintain the same orientation over intervals of about one-third of a band period. After that they can often be seen to split into two or three successors, accompanied by large changes of *c* axis orientation. This complicated morphology has still to be fully elucidated but it is clear that there is no support for the earlier view that they consist of arrays of continuously twisted helicoidal ribbons.

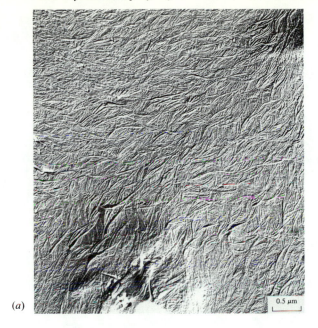

(a)

(b)

Fig. 4.33. View down the radius, *b,* of a banded polyethylene spherulite: (*a*) etched surface, (*b*) arrangement of the lamellae in part (*a*).

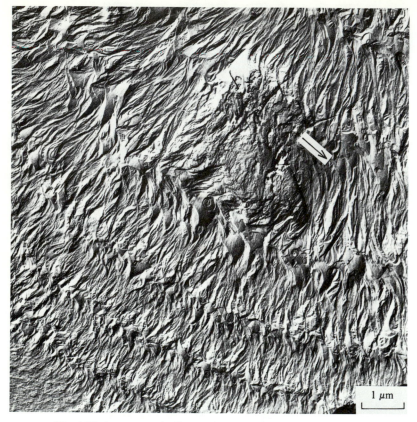

Fig. 4.34. An apparently diametral section of a similar banded polyethylene spherulite to that of Fig. 4.33. Note the splitting of lamellae (arrowed centre right). Etched surface. (From Bassett & Hodge, 1978*a*.)

Examination of further morphological detail shows that there is widespread fractionation within, and during the growth of, spherulites. It is inevitable, as dominant lamellae form first and ultimately penetrate to the limits of a spherulite, that any uncrystallized material must be located in the intervening channels as is showed by the results of quenching (Fig. 4.30). For the highest growth temperatures the difference in thickness between quenched and isothermally crystallized lamellae is large enough to permit selective extraction of the lower melting

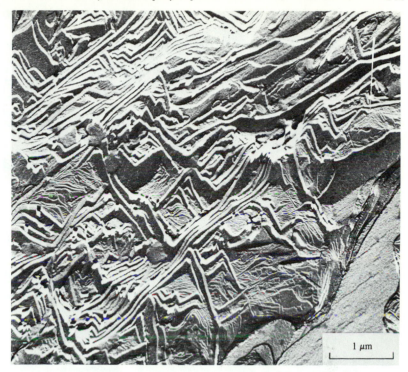

Fig. 4.35. Dominant planar sheets with subsidiary ridges of polyethylene ($\bar{M}_m = 3.1 \times 10^4$, $\bar{M}_m/\bar{M}_n = 1.3$) crystallized at 130.4 °C for 27 days plus thin lamellae formed on quenching. Etched surface. (From Bassett, Hodge & Olley, 1979.)

populations and determination of their molecular mass. As expected they contain predominantly the shortest molecules of the polymer. Moreover, one can obtain textures with dominant planar sheets separated by subsidiary ridges (Fig. 4.35). As the conditions to form a ridged population are rather specific (e.g. molecular mass of $\sim 3 \times 10^4$ at 130 °C) the subsidiary lamellae in Fig. 4.35 can be inferred to contain shorter molecules than the surrounding dominant sheets. This has major implications, as it implies a potential variation of properties from point to point within spherulites.

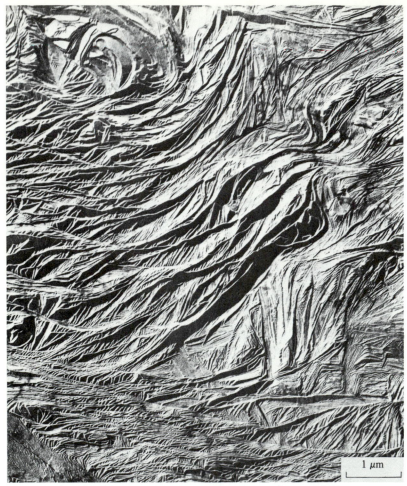

1 μm

Fig. 4.36. Microstructure of a commercial polyethylene (Rigidex 50) crystallized at 119.8 °C which contains S-shaped, planar and ridged lamellae. Etched surface. (From Bassett, Hodge & Olley, 1979.)

Similar morphological patterns are found in other polyethylene spherulites suggesting that similar processes have been operative, although for the more rapidly grown textures the inability to quench further is likely to make this difficult to confirm directly. Fig. 4.36 is just such an example. It shows the whole polymer Rigidex 50 crystallized rapidly at 119.8 °C and yet all three types of dominant sheets, ridged, planar and S-shaped are located side by side. The sample is far from homogeneous at the lamellar level. This is true even of quenched polyethylene which in addition to dominant Ss and their subsidiary platelets often contains small regions of ridged crystal, probably containing particularly short molecules. It suggests that the crystallization processes retain their subtlety even to very rapid rates of growth, and are more than a mere retention of molecular conformations roughly as they were in the melt.

Finally, one needs to consider the fine-scale morphologies observed in relation to the theory of Keith & Padden (1973) (Section 2.2). Qualitatively there is general accord, with clear evidence of lateral segregation of slower crystallizing species. Now, however, there is the possibility of measuring the lateral scale directly as the distance between locations of segregated species, i.e. subsidiary lamellae. As Keith and Padden have proposed, this can be seen to be approximately (i.e. within a factor of 3 or 4) the width of dominant lamellae. Fig. 4.37 shows a plot of laminar widths against molecular mass for conditions embracing changes by orders of magnitude in the growth rate and thus in δ. The lamellar widths, however, vary much less than δ and remain within a factor of three of 3 μm throughout. It is too soon to judge the significance of these very recent measurements for a theory which has been successful in accounting for optical spherulitic textures for more than a decade. They may, however, serve as an example of how essential it is in morphological research to re-examine even the most basic issues with new techniques as they become available.

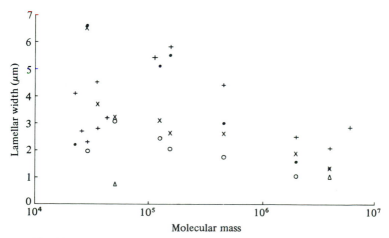

Fig. 4.37. Lamellar widths observed in polyethylene spherulites as a function of molecular mass: + 130; ● 128; × 125 ○ 120 °C and △ quenched from the melt. (From Bassett, Hodge & Olley, 1979.)

4.6   Further reading

Fuller details of the three-dimensional development of polymer crystals than are presented here may be found for various polymers as under: polyethylene, Bassett, Keller & Mitsuhashi (1963); polypropylene, Khoury (1966); poly (4–methylpentene–1), Khoury & Barnes (1972); polyoxymethylene, Khoury & Barnes (1974); polychlorotrifluoroethylene, Barnes & Khoury (1974).

5 Processes in crystallization and annealing

The morphological record contains information not only on the composition of a sample, but also on the nature of the treatments to which it has been subjected. For example, a polymeric specimen's lamellar thickness will point to its crystallization temperature, with the crystal habit indicative of the range of its molecular masses. In addition there are features which are pertinent to the character of the molecular processes involved in, for example, crystallization and annealing. These are the theme of this chapter.

5.1 Secondary nucleation

Among the most important observations is one of the earliest, showing that solution-grown polymer crystals continue growing with a changed thickness when their crystallization temperature is altered. Though present in some of the earliest micrographs of single crystals, this feature took some little while to be appreciated for what it was. If a commercial whole polymer of linear polyethylene is crystallized from 0.01% solution in xylene at 90 °C for 24 h and then cooled, one finds a population of truncated lozenges, probably with the appearance of doubled edges owing to a narrow external border, somewhat in the manner of Fig. 3.9. On the other hand, cooling after 12 h produces large crystals centred on small truncated lozenges as in Fig. 4.23. In both cases crystallization has occurred in two stages, isothermally at 90 °C, then on cooling, as may be confirmed using hot filtration.

Apparatus, such as that depicted in Fig. 5.1, allows isothermal crystallization to be conducted in the uppermost vessel for a chosen time, until the tap is opened, when the liquid drains away leaving a jelly of crystals on the filter. If some of these are redispersed in fresh xylene, phase-contrast or electron microscopy establishes that they consist of uniformly thick lamellae, truncated lozenges in the circumstances cited, with a single sharp edge. Isothermal growth from solution thus produces a lamella of constant thickness, a conclusion confirmed by small-angle X-ray measurements on sedimented crystal mats. The three to four orders of reflection observed for polyethylene crystals lead to estimates of a few per cent for the standard deviation of thicknesses around the mean.

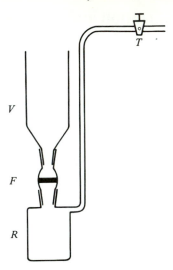

Fig. 5.1. Schematic of a hot-filtration apparatus. Crystallization occurs in the vessel *V* until the tap *T* is opened when the liquid flows into the receiver *R* leaving crystals behind on the filter *F*.

The filtrate will usually also contain crystals once it has returned to ambient conditions. These will be of lower average molecular mass and are likely to have formed on cooling. When the first small-angle X-ray measurements on polyethylene crystals grown from xylene at 85 and 90 °C were made, two different spacings were found. Hot filtration, of the type described, was used to demonstrate that only the higher appeared for the initial precipitate and only the lower for the filtrate. Moreover, the higher was constant for constant crystallization temperature while the lower varied with the conditions of cooling. This establishes the relationship to the two stages of growth, but the significance of the two X-ray long periods in unfiltered preparations is only that there are two thicknesses of crystal present. Microscopy is needed to demonstrate that both may occur within one lamella. Controlled experiments involving growth at two successive temperatures, either increasing or decreasing, have since shown quite clearly that growth after the temperature jump is at a correspondingly higher or lower value than the original layer. This is highly significant. It negates the early hypothesis that the thickness of a solution-grown polymer lamella is invariant and equal to that of its primary nucleus. Instead, the thickness is manifestly determined, subject to prevailing conditions, at the growing edge, i.e. by secondary nucleation.

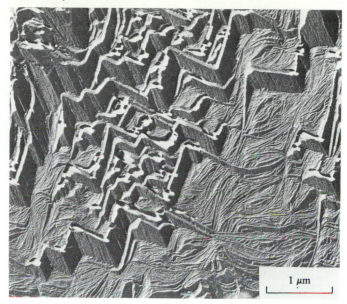

Fig. 5.2. A linear polyethylene with $\bar{M}_m = 21\,000$, $\bar{M}_m/\bar{M}_n = 1.7$ which was crystallized at 130.4 °C for 27 days and then quenched. The ridged sheets continue to grow sideways at smaller thicknesses but note the absence of a sharp dimensional change. Etched surface. (From Bassett, Hodge & Olley, 1979.)

There is, however, a major and still unexplained difference between melt-grown and solution-grown crystals in this respect. Whereas a downward step in growth temperature produces an electron microscopically sharp (to, say, 2 nm) boundary between old and new areas for solution growth, this is not true from the melt. Fig. 5.2 shows polyethylene lamellae which have suffered a fall in growth temperature, sufficient to change their habit from ridged to (half) S-shaped sheets but they reveal no clear internal boundary between the two portions. Instead, there is a gradual reduction in thickness. In one sense this is merely inconvenient since there is no step to delineate the growth front at a particular time, as may conveniently be employed in studies of growth rates from solution. More fundamentally it appears to show that, whereas molecular conformations from solution growth remain, except for minor readjustments, essentially as first laid down, there is considerable reorganization behind the growing edge during solidification from the melt.

This idea was first mooted following examination of the thick polyethylene lamellae produced anabarically at ~5 kbar (0.5 GPa). These frequently show tapering edges in fracture surfaces due to the

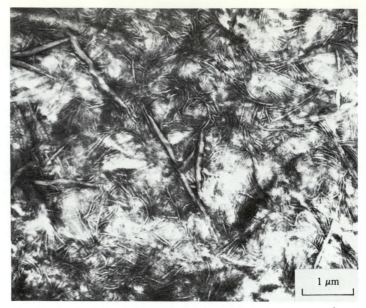

1 μm

Fig. 5.3. A mixed sample in which linear polyethylene ($M_m > 2 \times 10^6$) was crystallized at 2.30 kbar forming a few anabaric lamellae. These have continued to grow as thinner lamellae via a tapering transition zone. Chlorosulphonated sample. (From Hodge & Bassett, 1977.)

growing edge being thinner than the interior of the crystal. One can see this directly while watching individual, ~ 1 μm thick, crystals propagate in the high-pressure diamond-anvil cell mounted on a polarizing microscope (Fig. 7.11). A particularly effective demonstration is when, during growth in the region of 3 kbar (depending on molecular mass), polyethylene starts precipitating as the high-pressure disordered hexagonal phase but continues as the orthorhombic form. This leaves a curved silhouette separating the two regions of growth, Fig. 5.3, which may well be a reasonable representation of the *in situ* growth profile. The final lamellar thickness is clearly only established a distance, comparable with the thickness, behind the growing edge. As there is no good reason to regard anabaric crystallization of polyethylene as anomalous in this respect – similar profiles have since been observed, in suitable circumstances, in PTFE and in orthorhombic polyethylene grown *in vacuo* – it is probable that a tapered profile is a general feature of melt growth but particularly prevalent for longer molecules. Indeed, to explain their relatively high melting points, it has been found necessary to assume that the thickness of melt-crystallized lamellae increases by a factor γ, often ~ 3, over that nucleated at the surface (Section 6.1.5). So far this has been purely an empirical matter, but a fundamental justification is required to place understanding of melt crystallization on a firm basis rather than, as at present, on an arguable analogy with solution growth.

5.2   Fractionation

Reference has been made on several occasions to the fact that heating a suspension of truncated polyethylene lozenges in xylene to, say, 94 °C dissolves only their edges, leaving them surviving but smaller. The implication is that the outer regions are more soluble than the inner. It has been shown that this is due to a systematic distribution of molecular lengths across a layer. Longer molecules tend to precipitate earlier and shorter ones later so that if all lamellae nucleate simultaneously, as in self-seeding, the shorter molecules will be placed predominantly towards the periphery of crystals. Experiments involving hot filtration confirm that this is so.

The case of melt-crystallized polymers becomes more subtle. The three-dimensional structure of a spherulite with its skeleton of dominant lamellae will place later-crystallizing species within the interstices rather than at the edges of the early layers. In the example of Fig. 4.35 for crystallization at 130 °C, molecular masses of $\sim 3 \times 10^4$ crystallize after, and are placed between, planar lamellae consisting of longer molecules. They are also surrounded by lamellae which only solidified on quenching and contain the shortest molecules. This trend with molecular mass is common in the mass ranges usually encountered. There is not, however, a monotonic increase of crystallization rate with molecular length, but rather a temperature-dependent maximum (Fig. 5.4). In other cir-

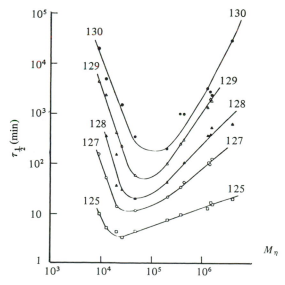

Fig. 5.4. Logarithm of crystallization half-time (in min) plotted against the logarithm of molecular mass for the five temperatures marked. (From Mandelkern, 1968.)

cumstances, and especially at lesser crystallization temperatures, one could, accordingly, expect to find longer molecules in the subsidiary lamellae though such a situation has still to be uncovered. Nevertheless, the placement of different molecular species in specific sites within the texture, as exemplified by Fig. 4.35, is likely to affect important properties such as stress-cracking resistance or embrittlement due to excessive concentrations of short molecules.

So far, fractionation has only been considered in terms of molecular length. It can occur for other reasons, notably short-chain branching. Just as the melting point of an infinite crystal decreases for lower molecular mass, so it does for higher branch concentrations. Moreover, it is frequently the case that, in a polydisperse randomly branched copolymer, shorter molecules tend to be more highly branched. Furthermore, there is also the probability, even for random branching, that side groups will be unevenly distributed along a molecule creating the possibility that branch-free segments may be incorporated within lamellae and portions of concentrated branching within the same molecule rejected to interlamellar regions. This certainly appears to be the case for branched polyethylene where, in particular circumstances, only methyl side groups have been found to be included within the lattice while ethyl and longer branches are excluded. With copolymers, therefore, fractionation can be considerably more complex than for linear macromolecules. In Fig. 5.5 is an appropriate example where there are four populations in the melting endotherm (Fig. 5.5a) and lamellae of four corresponding thicknesses, identifiable at four different kinds of location in the spherulitic morphology (Fig. 5.5b). The sample is a particle form copolymer of polyethylene, with 4.7 butyl branches per 1000 carbon atoms, crystallized by cooling from the melt at 15 Kh^{-1} and 5 kbar. Fractionation is particularly apparent in such anabaric systems and the range of thicknesses allows the lower melting populations to be dissolved selectively and their characteristics determined. In this instance the sequence in which crystals deposited is clear from their geometrical arrangement. First were the long, thickest layers then the thick infilling lamellae, both probably forming initially as the high-pressure (anabaric) phase. These contain the longest and most linear molecules as well as sufficiently long linear portions of branched molecules. The third family is repetitively ridged which identifies it as having crystallized as the orthorhombic structure under regime I conditions. Finally, the enveloping matrix of the thinnest lamellae will contain the most defective crystallizable mole-

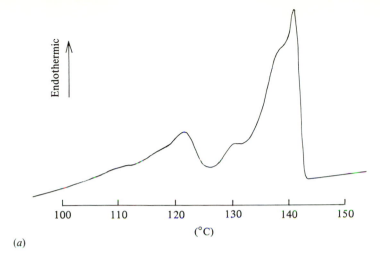

(a)

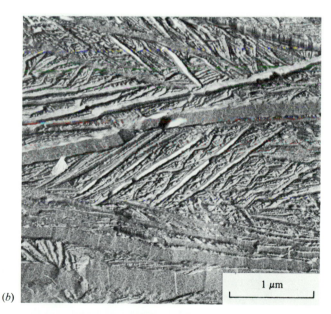

(b)

Fig. 5.5. (a) Melting endotherm with four resolved peaks, and (b) lamellar morphology with four corresponding thicknesses at different types of location within the morphology. The sample is polyethylene with 4.7 butyl branches per 1000 carbon atoms crystallized at 5 kbar while cooling at 0.05 K min^{-1} from the melt.

cules. One thus has the potential, in such systems, of studying the variation of properties and behaviour between specific components of the structure in relation to their differing molecular characteristics. Such information is likely to be particularly relevant, for example, to establishing mechanisms of failure.

Both kinetic and thermodynamic factors are involved in fractionation, the two being related by the theory of rate processes. The varying times at which different molecular species crystallize are evident both in macroscopic kinetics and, microscopically, from the detailed morphologies. In considering kinetics one must distinguish between the time at which crystallization occurs (to some finite extent such as 1, 10 or 50% development of crystallinity) and the time span for the process to proceed from, say, 10 to 90% crystallinity. Both these intervals are functions of molecular mass and its distribution (polydispersity). In general terms, fractionated polymers of moderate molecular mass crystallize at the shortest times, with both very low and very high molecular mass fractions taking longer (Fig. 5.4). This, however, can tend to oversimplify what is actually occurring. Comparisons of the time span for crystallization show that this is greater for a whole polymer than for fractions prepared from it and is an indication that internal fractionation is involved in crystallization. The extent to which fractionation occurs, however, has become more and more appreciated in recent years and, as the micromorphology shows, can be both considerable and elaborate.

For solution growth, although fractionation does occur, with shorter molecules concentrated towards the edges of lamellae, the lamellar thickness is practically invariant with molecular mass. In consequence fractionation was, to a very large extent, ignored in early work; even the molecular masses of the polymers themselves were often tacitly overlooked, not least because the means of measurement were rarely available. For melt crystallization the effects of varying molecular mass are much more evident (e.g. Fig. 5.6), and it is the greater study of these systems which has progressively revealed how widespread and important fractionation is.

One may distinguish two categories: fractionation due to the inability of certain, usually low molecular mass and/or highly branched species, to crystallize at the chosen temperature on one hand and time-dependent segregation on the other. The division is inevitably blurred according to the time scale of the experiments but may be illustrated for polyethylene crystallizing at 130 °C *in vacuo*. Under these conditions fractions of molecular mass $\sim 10^5$ crystallize completely within about 1 week; those of $\sim 4 \times 10^4$ require about 3 weeks while masses below

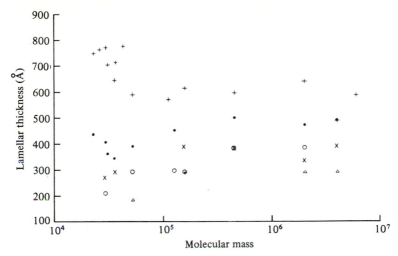

Fig. 5.6. The lamellar thickness, measured on etched cut surfaces, as a function of molecular mass for linear polyethylene fractions crystallized from the melt at different temperatures: + 130 °C; ● 128 °C; × 125 °C; ○120 °C and Δ quenched into ice-water.

~2 × 10⁴ are unlikely to solidify within this period. The fractionation in fig. 5.2 is, therefore, of the first category where the quenched background was effectively unable to crystallize. In Fig. 4.35, on the other hand, one sees also the planar and ridged sheets resulting from slower precipitation of shorter molecules. A similar situation prevails at 5 kbar where molecules < ~ 100 nm long are unable to form the anabaric phase and can only crystallize on cooling. Slightly longer molecules can form this phase in reasonable times, but only at the higher supercoolings, so that both categories of fractionation are found following cooling sequences (Fig. 5.5). In both examples cited the solidification of whole polymer shows a sequence similar to the sum of its constituent fractions with the longer molecules precipitating first (into dominant lamellae of appropriate shape), followed by successively lower molecular masses. One may conclude that crystallization rate is not controlled by primary nucleation otherwise crystallization would proceed rapidly to completion once this had occurred. Instead, one must needs look to the growth front and specifically to the processes of attachment and detachment of molecular segments to the developing crystal. One expects these to be governed by the thermodynamic functions of the two phases and, accordingly, to vary with molecular mass. Considering the crystallization of polyethylene from dilute solution, Sadler (1971) has shown experimentally that the fractionation which results is close to that expected if the lamellar crystals were actually in equilibrium with the solution, on the assumption that molecules could redistribute within laminae after deposition. It appears that, for solutions, the separation achieved by transport across the interface is similar to the ideal thermodynamic distribution, kinetic factors notwithstanding.

Fractionation is inevitable if the composition of a growing crystal differs from that of its environment. The enhanced precipitation of any one molecular species will change the constitution of the remaining melt or solution and thereby affect subsequent crystallization. It may, for example, decrease the supercooling of remaining species with respect to the melt and their crystallization rates correspondingly. Any such effect is, however, easily offset by cooling the melt. In this way, and particularly by slow cooling, a sequence of fractionated populations is likely to result, in the manner of Fig. 5.5 and 4.35.

5.3   Annealing behaviour

Annealing implies heating a solid to temperatures approaching its melting point. This activates internal mobility and promotes greater stability by, for example, the elimination of stresses or defects in a general movement towards the thermodynamic equilibrium condition. A glass blower will anneal his creations to remove or reduce the strains of fabrication. A cold-worked metal undergoes polygonization on annealing as its dislocations move and congregate in arrays of lower energy. Similar effects are found in polymers plus phenomena associated with the metastability of the lamellar crystals linked to their lowered melting points.

Reorganization of mats of solution-grown polymer crystals begins once they are raised above their original crystallization temperature. This is shown by sensitive adiabatic calorimetry which records the evolution of small quantities of heat (of order $Jg^{-1}h^{-1}$) once this temperature is exceeded. The concomitant increase in crystal perfection could, for example, involve the easing of strains due to crystal collapse on sedimentation or anisotropic thermal contraction. Soon, however, it becomes clear that the fold surfaces are becoming more regular. Annealing a mat of truncated polyethylene lozenges, precipitated at 90 °C, at higher temperatures shows a diminution of long period by ~ 1 nm from the original 15 nm. The beginning of this decrease has been recorded after 30 min at 95 °C and the reduction is complete after 15 h at 97 °C. At the same time X-rays show that the tilt of the c axis with respect to the lamella rises from ~ 30 ° appropriate to {(312)(110)} sectors to ~ 36 °.

The two values are consistent with the interfold length remaining constant, while lamellae become thinner because of the greater molecular inclination. The average separation of folds in the fold surface must increase accordingly (Fig. 3.16). This would be expected because, in a more ordered surface, with folds brought more nearly into one plane by the evening up of stem lengths, greater interaction between neighbouring folds would require more surface area for their accommodation.

The slope of these improved surfaces is near to {623} but attribution of such indices would have little significance. There could not be a unique fold structure or packing for this inclination, in agreement with the view adopted previously (Section 3.2.4). The most reasonable opinion appears to be that the fundamental asymmetries of folds lead necessarily to oblique habits but that the extent of the slope depends upon the degree of fold interactions in a particular case. It is entirely consistent with this view that, within a given population, experiment shows the variation in tilt to be less than between different populations formed on different occasions when, presumably, small variations in folding arrangements have resulted.

It is also pertinent that, if the same truncated lozenges are heated suspended in xylene to $\sim 94\,°C$, they also adopt more sloping habits. The evidence is that, if shortly afterwards the crystals are sedimented on a substrate, they show internal lines corresponding to collapse of surfaces near $\{(311)(110)\}$ (Fig. 4.14). These features are transient, disappearing within hours if the crystal suspension is stored at room temperature. They cannot, therefore, reflect permanent changes but probably relate to the packing requirements of folds swollen with hot xylene close to the dissolution temperature.

It is generally the case that fold surface regions become less dense, and reversibly so, with increasing temperature; this may or may not cause an increase in long period. The intensity of low-angle X-ray or neutron scattering rises with temperature showing, to a first approximation, that the integrated density deficit of the interlamellar zones has increased. In principle this could occur at constant thickness by the elimination of crystal defects with related changes in density of internal and surface regions. In practice the overall density tends to fall requiring an increase in long period to conserve mass; moreover, the interlamellar regions tend to grow at the expense of the interior. With an appropriate textural model the precise quantities can be evaluated by comparison of predicted and observed scattering curves. There are also other indicators, of which Fig. 5.7 is one. This shows a thin melt-crystallized film of linear polyethylene photographed slightly out of focus in the transmission electron microscope to bring a degree of phase contrast to the image.

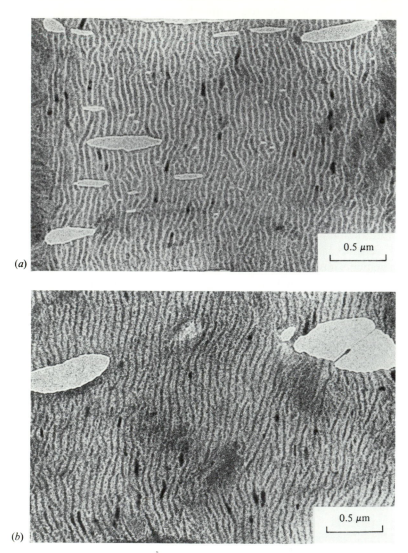

(a)

(b)

Fig. 5.7. Bright-field defocused electron micrographs of polyethylene film taken at room temperature: (a) was annealed for 12 h at 129 °C; (b) was crosslinked at 129 °C with ~5 Mrad of 100 kV electrons to fix the morphology. The bright lines correspond to 'amorphous' regions which are ~12 nm thick in (a) but 30 nm in (b) which corresponds to the high-temperature condition. (From Peterman & Gleiter, 1976.)

The view is of lamellae sideways on with lighter strips corresponding to the surface regions. Particularly interesting is the observation that these lighter strips are thicker the hotter the specimen is inside the microscope, in accordance with the statement above.

There are also irreversible increases in long period which were first discovered in work of Statton & Geil (1960) on solution-grown lamellae of linear polyethylene. They are part of phenomena now known collectively as lamellar thickening which embrace not only annealing behaviour but also isothermal crystallization from the melt. The fundamental cause is that polymer lamellae are metastable because they are thin. If their thickness increases the relative energetic cost of their fold surfaces will fall and bring greater stability. The means by which these changes are effected involves, at least at higher annealing temperatures, a mechanism of melting the least stable parts of a sample and recrystallizing them at a higher thickness around the nuclei provided by the remainder. This is, in effect, self-seeded melt crystallization which progressively changes the character of initially solution-grown aggregates. This is implicit in Fig. 3.21 where the increased slope of the density against inverse lamellar thickness plot shows that the density deficit of the interlamellar regions has increased irreversibly with the lamellar thickness, but remains constant at a given temperature even though the lamellar thickness then increases linearly with the logarithm of the annealing time. At lower annealing temperatures, which may appear to lie below the melting range, the evidence for melting is less obvious. In these conditions it has often been supposed that a different molecular mechanism obtains known as solid-state thickening. Discussion of these matters requires the introduction of further information.

The observations of Statton and Geil were that at temperatures of 110 °C and above solution-grown polyethylene lamellae suffered irreversible but time-dependent increases of thickness. Their samples contained 10.4 nm thick laminae, grown simply by being cooled to room temperature. Conversely, the 13 to 15 nm thick truncated lozenges grown at 85 and 90 °C show only improved fold surface packing, without thickening at such temperatures. For these, temperatures of ~ 125 °C and above are required for the long period to increase.† Correspondingly, lamellae thinner than 10.4 nm start to thicken below 110 °C. This is also true of polyethylene quenched from the melt, where the change is almost certainly confined to just the lowest melting part of the population. Samples containing two populations of solution-grown

---

† At these higher temperatures there is also molecular reorientation involving rotation of the unit cells around the *b* axis for truncated lozenges, but around the *a* axis for thinner lamellae.

lamellae show directly that those with the thinner long period begin to thicken first independently of the other (cf. Fig. 5.16). There is quite reasonable agreement between the onset of long period increase and the melting point expected for lamellae of the corresponding thickness according to equation (3.1)

$$T_m = T_m^\circ \left(1 - \frac{2\sigma_e}{\Delta h l} \right)$$

which for $\sigma_e \sim 100 \, \mathrm{mJ \, m^{-2}}$ predicts a melting point depression of $\sim 30 \, \mathrm{K}$ for a 10 nm thick lamella, etc.

There is a clear temporary loss of order at the higher annealing temperatures indicated by a transient substantial reduction in X-ray crystallinity (i.e. more intense diffuse but weaker sharp wide-angle reflexions), lowered birefringence and a fall in density. The centres of mass of molecules are also found to move apart from neutron scattering data. The higher the annealing temperature, the longer is the recovery but this is always faster than the time for isothermal crystallization from the unperturbed melt at the same temperature (Fig. 5.8). These observations leave little doubt that, at least at these high temperatures, annealing behaviour is indeed a consequence of melting followed by recrystallization around residual crystallites acting as seeds. The time-dependent increases of the long period probably reflect both fractional recrystallization and thermally activated improvement of crystallinity.

The mechanism cited leads to the expectation that the original texture will influence the nature of the product. This is so, with the most obvious

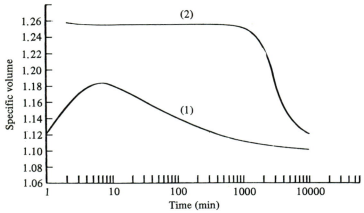

Fig. 5.8. Dilatometric isotherms measured at 130 °C for Marlex 6000 linear polyethylene: (1) while annealing, and (2) for crystallization from the melt. (From Matsuoka, 1962.)

manifestation the maintained chain-axis orientation. For example, carefully sedimented mats of solution-grown lamellae are likely to have *c* preferentially normal to the specimen plane. This generally persists, though with some randomization, on annealing to the melting point† despite very substantial lamellar thickening. Especially striking is the maintained (though weakened) ring pattern in banded polyethylene spherulites after annealing from the orthorhombic into the disordered hexagonal high-pressure phase at 5 kbar, a transformation which increases the average lamellar thickness tenfold from 20 to 200 nm.

Annealing of polyethylene at high pressures is particularly instructive regarding the nature of lamellar thickening because of the large changes concerned. That melting is involved is strongly supported by the fact that annealing in the orthorhombic phase at 5 kbar produces thicknesses to ~50 nm for the linear polymer, only a little below those produced by direct crystallization of this phase from the melt, whereas anabaric annealing (in the high-pressure hexagonal form) gives much thicker lamellae (~200 nm), approaching the 400+ nm produced by direct crystallization of this phase from the melt. Orientation disappears at the melting point of the hexagonal phase‡ and taking to still higher temperatures results in typical melt-crystallized products, with greater thicknesses than produced by annealing.

It appears to be a general result in linear polymers that annealing at a chosen temperature gives a lower lamellar thickness than does crystallization at the same temperature. There are at least two factors responsible for this difference. The first is that melting only part of the sample and then recrystallizing it at the annealing temperature must always give a lower average thickness than a sample all of which has been solidified at this same temperature. This implicitly assumes that the lamellar thickness is a single-valued function of temperature (extrapolated to $\log t = 0$). Annealing results cannot, however, be fully accounted for on this basis and are influenced by the initial morphology.

It can be shown quite clearly that crystallizing polyethylene in restricted volumes leads to lower lamellar thicknesses than result from the unconstrained melt. At high pressures reductions greater than an order of magnitude can result. Crystallization of linear polyethylene from the bulk as the hexagonal phase leads to lamellar thicknesses of ~400 nm at 5 kbar whereas recrystallizing solution-grown lamellae on a substrate in

---

† The rotations around *a* and *b* for polyethylene very close to the melting point mentioned in the previous footnote are an exception to this behaviour.

‡ Except for polymer of very high molecular mass (~10^6) whose large viscosity allows the chain orientation to stay, in some degree, at higher temperatures for experiments of an hour's duration.

Fig. 5.9. The optical texture, between crossed polars, of Rigidex 9 linear polyethylene after heating for 15 min at 5 kbar and 254°C. Note the reduced lamellar sizes in the upper half of the field. (From Bassett, Khalifa & Olley, 1976.)

these conditions produces thicknesses of only 40 nm or less despite clear evidence (from diffraction patterns) that the hexagonal phase was involved. Similar effects have long been suspected to occur with the orthorhombic structure at atmospheric pressure and have recently been identified. They probably involve the suppression of isothermal thickening and lead to thinner, tapered edges to lamellae as they fill in crevices in their surroundings. Similar effects may also be expected to occur on annealing and Fig. 5.9 shows that recrystallization within the constricted spaces between existing laminae produces layers of substantially lower thicknesses than occur in neighbouring unrestricted volumes. The extent of the reductions which result has not yet been quantified to more than the extent outlined but they allow a consistent qualitative picture of the changes occurring on annealing. This can be demonstrated by considering the melting endotherms of annealed samples.

The melting endotherms of crystalline polymers became widely studied following the introduction of commercial differential scanning calorimeters in the mid-sixties. These provide a plot of the power

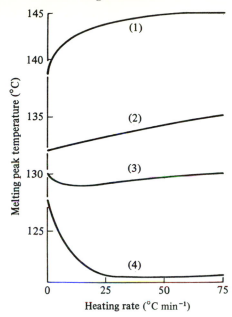

Fig. 5.10. Melting peak temperatures of the same linear polyethylene as a function of heating rate for different morphologies. (1) after anabaric crystallization; (2) slowly melt-crystallized; (3) quenched from the melt; and (4) solution-grown. (From Wunderlich, 1973.)

supplied to a sample to keep its nominal temperature rising at a constant rate. It was soon found, however, that the endotherms recorded differed substantially according to the rate of heating used. The variation of heating rate on the recorded melting endotherm of homogeneous samples of polyethylene lamellae is shown in Fig. 5.10 and is compounded of two opposing effects. Faster heating rates may tend to move the values of peaks, etc., to higher values even when inherent thermal lags in the apparatus have been corrected. This is superheating. At the same time there is reorganization within the sample during heating, of the kind found on annealing. With sufficiently fast heating rates (above 32 K min^{-1}, say) there may be insufficient time for this to be completed *en route* so that the melting endotherm then shows two peaks, one of reorganized material and one which will have comparatively little reorganization but may still not be completely unaltered. The balance between these two peaks can be shifted by heating rate and also, for example, by radiation crosslinking which slows reorganization and thereby favours retention of the lower peak. It is important, therefore, in morphological work to be aware of the complexity which is often hidden

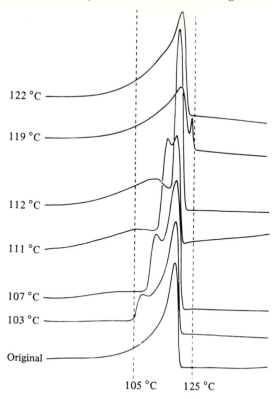

122 °C

119 °C

112 °C

111 °C

107 °C

103 °C

Original

105 °C   125 °C

Fig. 5.11. Melting endotherms of polyethylene with 7.5 ethyl branches per 1000 carbon atoms after annealing at 1 bar for 2 min at the marked temperatures.

in an apparently simple melting endotherm. Nevertheless, for suitable heating rates, usually ~ 10 K min^{-1}, a single lamellar population will give a single peak in the melting trace so that changes in the nature of populations produced by annealing (or other treatments) may be followed by the changes in DSC output.

As an instructive example, consider Fig. 5.11 which relates to a lightly branched polyethylene (with 7.5 ethyl groups per 1000 carbon atoms). This material when crystallized by quenching from the melt shows a broad melting endotherm peaking at 119 °C. Annealing at a low temperature within the endotherm, such as 103 °C, followed by quenching, changes the melting trace so that there is no longer any marked melting before 103 °C, but at this temperature the new curve rises to accommodate what remains of the old trace plus the annealed and recrystallized material whose melting has now been pushed above the annealing temperature. Evidently, the crystallinity of material which had melted below the annealing temperature has been improved and the constituent crystals have become more stable.

This is a trend which continues as the annealing temperature, T_a, rises through the melting curve, but T_a does not rise very far before some of the material melted out below T_a fails to recrystallize at T_a in the time available and only solidifies on cooling. This has happened at, for example, 111 °C, producing a new broad low peak in the endotherm at temperatures below that of the original. The reason for this is that molecules comprising the lowest melting tail (and latest crystallizing portion) of the original sample now tend to be more concentrated making it likely that the crystals they form will be more imperfect, with a lower melting point than before. This, therefore, adds a second effect of annealing and, as T_a increases, the two peaks due to (i) recrystallization at T_a and (ii) recrystallization on quenching, rise through the original melting endotherm until its presence is obliterated. At this stage a sharp high-temperature peak begins to emerge which eventually melts at a higher temperature than the original material. This must refer to the melting of more stable (in fact thicker) crystals than were able to form under the initial conditions and, as is suggested by only part of the sample being involved, also consists of the most linear molecules or sequences therein. Once T_a exceeds the high end of the initial melting range, quenching restores the original behaviour. However, holding at longer times at only slightly higher temperatures gives a still higher melting point of thicker lamellae. In this way the highest melting point materials can be obtained, exploiting the self nucleation of the partly molten polymer.

Annealing of this and other highly branched polyethylenes at high pressures yields still further information. So long as annealing continues to be of the orthorhombic solid the phenomena correspond to those described above at 1 bar but raised by ~ 108 K at 5 kbar (Fig. 5.12). Once the high-pressure disordered hexagonal phase is encountered, however, the reformed lamellae become thicker and melt some 10 K higher than previously. Nevertheless, the crystal thicknesses are still considerably below those given by the linear polymer. This is a consequence of branches longer than methyl being excluded from the crystals under these conditions. The observed thicknesses are limited by the available interbranch lengths of the random copolymer, taking fractionation into account. It has been shown, for instance, that in a polyethylene with 7.5 ethyl branches per 1000 carbon atoms the longer molecules have a smaller branching ratio, 3 per 1000 carbon atoms, and that the highest melting population produced on annealing, consisting

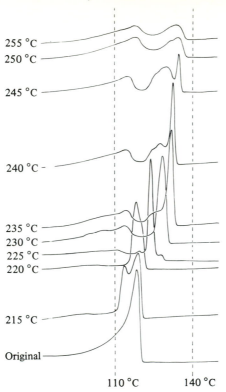

255 °C
250 °C
245 °C
240 °C
235 °C
230 °C
225 °C
220 °C
215 °C
Original

110 °C   140 °C

Fig. 5.12. Melting endotherms of the same polymer as in Fig. 5.11, but now after annealing at 5.35 kbar for 15 min at the marked temperatures.

of 15% of the sample, has a thickness of 55 nm in good agreement with the expectation from random statistics for this branch ratio (see also Section 7.2.3).

Such detailed evaluations provide very strong evidence, therefore, that the processes occurring when a polymer is annealed within its melting range are a consequence of its becoming partly molten and then recrystallizing within the surrounding matrix: this we term local melting and recrystallization. Much of the controversy surrounding the interpretation of annealing behaviour arises from phenomena at lower temperatures which are not obviously within the melting range.

It is not an easy matter, experimentally, to define the precise melting range of a crystalline polymer, especially in the early stages. This is further complicated because, even when melting has occurred, subsequent recrystallization can remove most of the heat evolved making

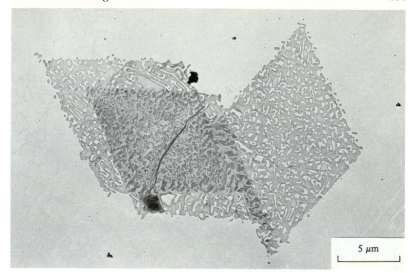

Fig. 5.13. Solution-grown polyethylene lamellae developing holes (but only in regions where layers did not overlap) after annealing on a substrate at 118 °C for 5 min.

the net output small. Particularly with conventional DSC equipment such modest effects are hard to detect with confidence. With adiabatic calorimetry heat evolution is detected but the source has still to be distinguished from the energy available from the elimination of defects. It is also likely that other phenomena customarily associated with melting, such as fluidity, may not be perceptible either. If the time scale of recrystallization is short, then the possible extent of flow must also be restricted. In one case, where there is a sharp onset of fluidity as the annealing temperature rises, the explanation is not a sudden change in mechanism to melting and recrystallization but reflects a more subtle change in the system. This is the high-pressure annealing of polydisperse whole polymer of linear polyethylene whose contacting pellets fuse at temperatures $> \sim 238$ °C at 5 kbar but remain separate if this figure is not attained. The reason is that under these conditions the system is beginning to undergo the transition from orthorhombic to disordered hexagonal phase, but that short molecules ($< \sim 100$ nm long) are unable to enter the new phase and, instead, these alone melt, flow and bond the pellets together.

Electron microscope observations show that solution-grown polymer monolayers annealed after sedimentation on a substrate develop holes in the range where lamellar thickening occurs but generally retain their precise crystallographic orientation after the change (Fig. 5.13). The holes are an artefact, only being developed in the presence of an intro-duced substrate and not when two polymer layers overlie. Such observa-tions, which were made shortly after the discovery of polymer single crystals, led rather naturally to the idea that the thickening was a property of the crystal and was proposed to be by some process of *solid-state thickening*.

This is by no means an implausible suggestion. It is well known that, in short-chain compounds, there are strong, rapid combined rotations and translations which allow a chain to move along the direction of its length within a lattice, particularly when associated with a neighbouring vacancy. This would be likely to be true also of polymers and, indeed, something of the sort may be responsible for the adjustment of folds in the improved packing of fold surfaces which develops in the early stages of annealing. A particular defect, the 'point dislocation' of Reneker (1962), has been proposed to facilitate the movement of a chain by passing the defect along it, rather than requiring a translation of the whole. This only involves a single chain whereas in local melting one would expect nearest neighbour stems, at least, to be perturbed. It has also been pointed out how particular patterns of folding molecules into ribbons, specifically by having an interstitial arrangement, would greatly facilitate a thickening process (Fig. 5.14). There is, nevertheless, no strong evidence that this actually is the mechanism responsible for thickening, but until recently there have been strong objections to regarding true melting as having taken place, despite the considerable circumstantial evidence in favour.

Morphological evidence shows that the onset of lamellar thickening is a sharply defined phenomenon. The early transformation of {100} sec-tors only in a truncated polyethylene lozenge (Fig. 3.20) already shows the sensitivity of the phenomenon. This is matched by the observation that higher layers in a multilayer crystal transform earliest, i.e. third

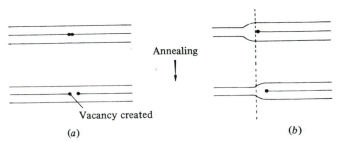

Fig. 5.14. A suggested mechanism whereby lamellar thickening is facilitated by the occurrence of an interstitial folded ribbon. The situation in part (*a*) is more costly of energy because of the creation of stem vacancies. (From Balta Calleja, Bassett & Keller, 1963.)

before second and second before first. While the overall surface/volume ratio increases for higher layers because of the contribution of the side surfaces giving them a lowered stability, thickening does, in fact, start in separate small areas in one layer (Fig. 5.15) and in thinner layers before thicker (Fig. 5.16). This shows that the local free enthalpy is the relevant quantity concerned in the initiation of the transformation, doubtless at suitable defects (Fig. 5.17*a*). Moreover, the observation that the first transformed regions lie along specific crystallographic directions – the *a* axis in Fig. 5.16 – suggests that the planes of lowest free energy have become exposed, with the further implication that a precise energy balance governs the onset of thickening.

There was, however, one effect which appeared to rule out a true melting in that the result of annealing at a given temperature depended on a sample's history. At that time it was believed, as is true for solution growth, that lamellae grown at a given temperature from the melt had a single-valued mean thickness. What is observed on annealing two samples of the same molecular mass, but grown from solution at two different temperatures to give two different X-ray long spacings, is that the smaller long period begins to increase at a lower temperature (Fig. 5.16) as would be expected because of the lower melting point and, surprisingly, that when both samples thicken at the same temperature the higher resulting long period is attained by the sample which, initially, had the thinner lamellae. On a simple melting plus recrystallization hypothesis they might have been expected to move towards the same value.

One possible inadequacy in the melting hypothesis is that molecular conformations might be constrained in some way, for example, by preserving the chain-axis direction in the molten state. This would be partial melting in the sense that not all degrees of freedom of the true melt would be taken up, and needs to be distinguished from the situation of a partly molten sample for which the term partial melting is widely misused. Any such phenomenon, however, appears to have little effect as judged from the behaviour of orthorhombic polyethylene with increasing pressure whose lamellar thickening curve is, to a first approximation, reducible to one of constant supercooling for various pressures. The melting point, T_m, will increase according to the Clausius–Clapeyron equation

$$\frac{\mathrm{d}T_m}{\mathrm{d}p} = \frac{\Delta v_f}{\Delta s_f}$$

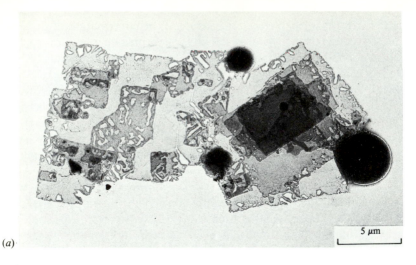

(a)

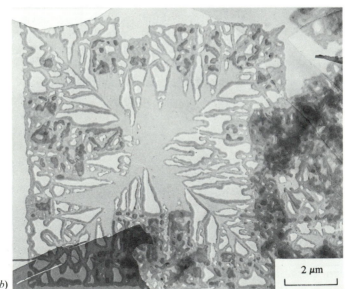

(b)

Fig. 5.15. Successive stages in the thickening of solution-grown lamellae of poly(4–methylpentene–1) after annealing on a substrate in air for 5 min (*a*) at 218 °C, (*b*) at 220 °C. Note in (*a*) the tendency to create holes parallel to ⟨110⟩ and to start thickening at or near crystal edges. The sector boundaries have become prominent in (*b*).

Fig. 5.16. Thickening beginning in solution-grown polyethylene lamellae annealed on a substrate. The thicker regions and holes tend to be parallel to *a*; they occur in the 12 nm border to the truncated lozenge but not its 15 nm interior.

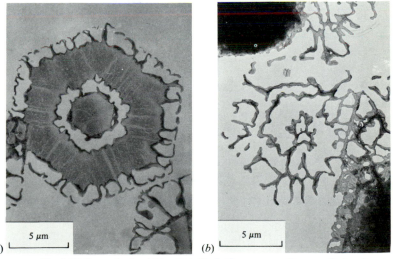

(a)   (b)

Fig. 5.17. Stages in the thickening of solution-grown polyoxymethylene lamellae after annealing on a substrate at ~200 °C for 5 min. In (a) thickening has started at the outer edge and round the internal seed crystal.

where Δv_f and Δs_f are the volume and entropy changes on fusion. This ratio would, in general, be expected to alter for partial melting; that the lamellar thickening curve moves with the melting point is, therefore, evidence in favour of melting being involved.

It is now known that crystallization from the melt does not give a thickness depending only on crystallization temperature: the volume available for a crystal to form is also a factor (Fig. 5.9). By imposing constraints on a crystallizing sample the lamellar thickness actually produced can be substantially depressed. This was observed first for polyethylene crystallizing as the high-pressure disordered hexagonal phase and has now also been observed for ordinary growth of the orthorhombic form. Put crudely, if only a restricted volume is available crystallization may still occur but the lamellar thickness will be reduced, presumably because time-dependent thickening after crystallization is suppressed. In this way an outline solution of the variation of long period with sample history on annealing appears possible. At a given (moderate) annealing temperature more of the sample with thinner lamellae will melt so that recrystallization will, on average, be in bigger spaces within the crystalline matrix. This less-constrained situation will lead to less depression of long period, i.e. to greater observed thicknesses.

The mechanisms of lamellar thickening on annealing may thus be as follows. At a sufficiently high temperature defects will initiate melting in small, preferably thinner, regions of lamellae (Figs 5.15–17), with lower molecular mass material affected first. Recrystallization, at a higher and stabler thickness, will follow rapidly (because of the relatively high supercoolings involved). Higher temperatures will involve more and more of the sample leading to still greater thicknesses but appearing more slowly with the time available for flow to occur increasing correspondingly.

This appears a reasonable hypothesis but has still to be proved. So far as the author is aware there are no clearly contradictory observations but many which it can explain. Ones additional to those already mentioned include the gradual change towards melt-crystallized characteristics which NMR shows occurs as solution-crystallized polymers are annealed, i.e. the mobile (disordered) fraction increases progressively. This changeover is also reflected in properties. Particularly noteworthy in this respect are the experiments of Blackadder & Lewell (1970a,b,c). These authors carefully sedimented solution-grown polyethylene lamellae into macroscopic specimens large enough to prepare dumb-bell shaped specimens 2.5 cm long for tensile testing. The several properties measured, melting point, Young's modulus and density, show gradual transformation from solution-crystallized character (brittleness, high density, etc.) to those typical of melt-crystallized (ductility, lowered density) (Fig. 5.18).

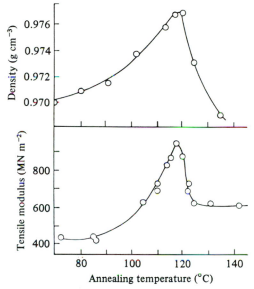

Fig. 5.18. Changes of density and tensile modulus for aggregates of solution-grown polyethylene crystals following annealing for 18 h at various temperatures. The final point for tensile modulus is that for a bulk specimen of the same density (0.972 g cm^{-3}). (From Blackadder & Lewell, 1970c.)

So far as melt-crystallized polymer is concerned there have been relatively few electron microscope observations, because of the problems of specimen preparation already recounted. However, even when fracture-surface replication was the only technique readily available, it was found that specimens which had been annealed or held for long times at the crystallization temperature showed better resolved lamellae than those crystallized from the melt for short times. This is, doubtless, a reflection of the improved order which has resulted. Most amenable to fracture-surface replication are the thick lamellae of anabaric polyethylenes. Detailed studies have been made of this system, particularly with a view to making oriented specimens with high chain-extension, it being known that the overall *c* axis orientation of a specimen drawn at 1 bar is preserved on annealing, in both orthorhombic and high-pressure hexa-

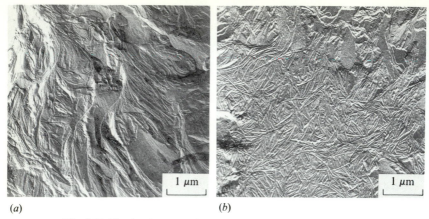

(a)   (b)

Fig. 5.19. The development of texture with annealing time in a melt-crystallized linear polyethylene (Rigidex 9). The original (a) was cooled from 160 °C to 110 °C at ~8 K min⁻¹ and then quenched in ice-water. Annealing was at 130 °C for (b) 0.5 h, (c) 3 h and (d) 95.5 h followed by quenching.

gonal phases.† In this system the continual increase in lamellar size (in all dimensions) becomes very clear once annealing occurs in the hexagonal phase. The density and melting point also increase appropriately. Annealing a melt-crystallized polymer under the same conditions gives similar results except that then the developing lamellae are constrained by the remnants of the original spherulitic ordering, such as the periodically rotating orientation associated with optical banding, which persists albeit in weakened form.

With the introduction of permanganic etching, details of the annealed morphology of all polyethylenes have become accessible. In the case of anabaric polyethylenes they confirm the fracture surface studies just mentioned. These reveal, however, a remarkable result when polyethylenes containing dominant S-shaped lamellae are annealed at 130 °C. This is that the dominant Ss are still clearly identifiable, though they may be doubled in thickness. The situation is illustrated in Fig. 5.19. The retention of individual lamellae in this way must imply that the thickening process (of local melting and recrystallization) has been restricted, at any one time, to volumes small compared to a lamella. Otherwise, the

† Triaxially oriented polyethylene specimens, i.e. with *a, b* and *c* axes mutually orthogonal, can be prepared by controlled relaxation consequent upon the annealing of drawn and rolled (or equivalently deformed) polyethylenes. This can be achieved for branched polymer at 1 bar but the best orientation is produced when the linear polymer is annealed in the orthorhombic phase at high pressure.

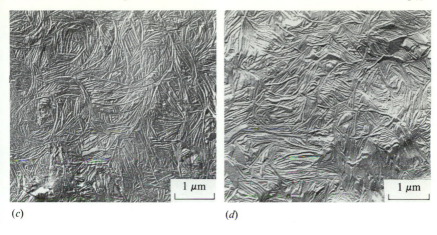

(*c*)   (*d*)

characteristic S-shaped profile would have disappeared and the newly
forming lamellae could be expected to retain the *c* axis orientation but to
have ridged or other profiles according to their molecular masses (Section 4.5.3). This does happen to some limited extent but, in typical
polyethylenes, only represents some small minority of the material.
There must, nevertheless, have been substantial addition of material to
an individual lamella as, despite the increase of thickness, there is no
marked alteration in width or length.

5.3.1   Annealing in suspension

The annealing of solution-grown lamellae while suspended in a solvent,
usually in concentrated form, was studied shortly after the discovery of
polymer single crystals. The temperatures at which dissolution begins,
for example 90–95 °C for polyethylene lozenges, are low compared to
the temperatures at which thickening occurs in the dried lamellae.
Nevertheless, there is limited thickening, but only at the edges and,
sometimes, inter-sector boundaries (Fig. 5.20). These locations make it
probable that thickening has proceeded through the agency of the
solvent, with cilia and other available chain sequences able to crystallize
at a greater thickness than the original lamella, because of the higher
temperature. A narrow thickened strip at the edge of a truncated lozenge

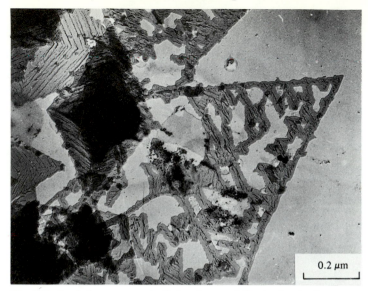

Fig. 5.20. A solution-grown polyethylene lozenge following heating in suspension to 94 °C. The interior has mostly dissolved except along sector boundaries.

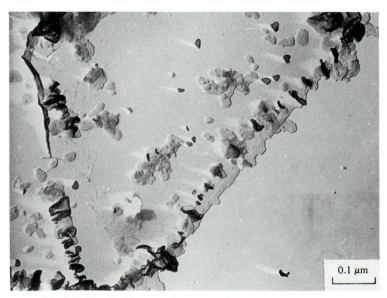

Fig. 5.21. Detail of the edge of a truncated polyethylene lozenge which had been heated, suspended in xylene, to 94 °C for 1 h then quenched showing the thickened outer edge. Note the structure of the shish kebab morphology in which lamellae have nucleated on a central thread.

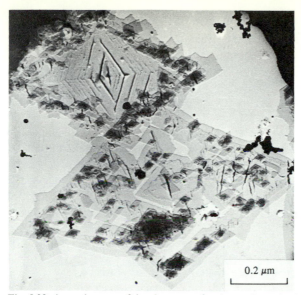

0.2 μm

Fig. 5.22. An early stage of development of a picture-frame habit in polyethylene after heating in suspension. The interior has begun to dissolve and the thickened exterior to grow profuse spiral growth terraces.

after ~1 hr annealing in xylene at ~94 °C is illustrated in Fig. 5.21. Note the tendency to form a jagged edge. This is relevant to a striking picture-frame habit often formed, especially for the thinner initial lamellae (Fig. 5.22). Once the edge has thickened, as just described, then it is more stable against dissolution than the interior which once the solvent gains sufficient access in some, possibly accidental, way may dissolve leaving only the exterior, plus other thickened regions, often including sector boundaries.† When the solution is cooled there will be growth on the thickened border and, because of its jagged profile, this is likely to lead to profuse thickening by spiral terraces, just as occurs in the similar circumstances of deposition on the {100} faces of truncated lozenges (Fig. 1.3). In this way the border becomes still more prominent (Fig. 5.22).

5.4   Further reading

Although there has been much work on annealing of polymers, this is an area where a firm consensus of opinion has yet to emerge. For further information on particular items the reader is probably best advised to turn, in the first instance, to the general reviews cited at the end of Chapter 1.

† This may suggest that these are special places where cilia, etc., may be especially long and is, perhaps, evidence in favour of molecules folding regularly along growth faces and having a zone of increased cilia concentration where sectors meet.

6   Theories of crystallization

The preceding chapters have indicated something of the complexity of polymer morphology. It is not unexpected, therefore, that at this still relatively early stage in the subject's development much of the understanding remains at a qualitative or semiquantitative level. There are, however, quantitative theories of the lamellar thickness and the crystallization rate which have met with considerable success and are widely accepted.

6.1   Theory of the lamellar thickness

Modern theories of the lamellar thickness are kinetic theories, that is they assume that the observed lamellar thickness is that which grows fastest and is not necessarily the most stable crystal which could have been grown. There have also been various equilibrium theories which have attempted to provide an explanation in terms of there being a minimum of free energy at a particular thickness but, by and large, these have not been generally accepted. On the contrary, it is recognized that, as with crystals of simpler solids, the shape of a real crystal is not the ideal equilibrium one but has been formed in response to kinetic factors. Put another way, and to paraphrase a comment by F. C. Frank (see Frank & Tosi, 1961), any minimum in free energy predicted on equilibrium grounds is likely to be broad and shallow especially in comparison with the free energies of crystallization.

The theory outlined below is a nucleation theory which assumes that a particular stage of nucleation is the rate-controlling step in the growth of polymer lamellae and links the observed lamellar thickness to that of the relevant nucleus. This type of theory immediately gives the correct sign of temperature dependence, with thicker lamellae forming the nearer the melting point is approached, for the reason that the size of a critical nucleus is governed by the need to offset the additional free energy of the surface by the reduction in free energy as the interior crystallizes. Whereas the surface terms vary only little with temperature, the free energy of crystallization decreases rapidly towards zero as the supercooling decreases so that larger and larger volumes (i.e. thicker and thicker lamellae) are needed to counteract the surface terms and produce a net reduction of free energy.

All the current theories have the premise that once molecules are added to a crystal subsequent readjustment is either zero or small. While this is thought to be the case for growth from solution, it is very doubtful that it also holds for growth from the melt where there is considerable lamellar thickening behind the growth front as has been discussed in Chapter 5. Nevertheless, the theories have been carried over to the melt case on an empirical basis. The earlier theories were formulated entirely in terms of infinitely long molecules for reasons whch will become evident. More recent developments have attempted to introduce finite molecular lengths.

6.1.1   Primary nucleation

The first nucleation theory assumed that the thickness of the primary nucleus was equal to the thickness of the lamella. Though this was soon shown to be in error, because the thickness of a lamella responds to changes in crystallization temperature, it still provides a useful introduction to the general concepts.

Consider a nucleus of thickness l containing v chains each a distance a apart in square array. (The actual geometry of the crystal structure only modifies the mathematics by constant factors.) The volume of the nucleus is thus $va^2 l$, the side surface area is $4la\sqrt{v}$ and the fold surface area $2va^2$. Then the excess free enthalpy (Gibb's function) due to the creation of the nucleus is

$$\Delta G = 4la\sqrt{v}\sigma + 2va^2\sigma_e - va^2 l\Delta f \qquad (6.1)$$

where σ is the side surface free enthalpy per unit area, σ_e that for the fold surface and $\Delta f \simeq \Delta h \Delta T / T_m^0$ is the free enthalpy per unit volume of crystal; Δh is the heat of fusion per unit volume at the equilibrium melting temperature T_m^0 and ΔT the supercooling. ΔG shows a maximum because the first two terms are always positive and only when v, l and Δf are sufficiently large can these be outweighed by the third term. The condition that the free enthalpy is a maximum defines the critical nucleus; its size is given by

$$\frac{\partial \Delta G}{\partial v} = \frac{\partial \Delta G}{\partial l} = 0 \qquad (6.2)$$

leading to

$$l^* = \frac{4\sigma_e}{\Delta f}$$

and

$$\Delta G^* = \frac{32\sigma^2\sigma_e}{(\Delta f)^2} \qquad (6.3)$$

According to classical nucleation theory the rate of formation of primary nuclei is

$$S \simeq \frac{NkT}{h} \exp - \frac{\Delta\phi}{kT} \exp - \frac{\Delta G^*}{kT} \qquad (6.4)$$

$$\simeq \frac{NkT}{h} \exp - \frac{\Delta\phi}{kT} \exp - \frac{32\sigma^2\sigma_e}{(\Delta f)^2 kT} \qquad (6.5)$$

where N is the number of molecules, i.e. Avogadro's number, kT/h is the frequency of thermally activated vibrations (k being Boltzmann's constant and h Planck's constant) while $\Delta\phi$ is the activation energy of transport for a molecule to cross the phase boundary, i.e. to reach the crystal surface. Although this theory is based upon wrong premises, it leads to expressions which are similar to those of more sophisticated theories and it is worth considering their main features. First of all the value of l^* is of the correct order. Expanding equation (6.3) by setting $\Delta f \simeq \Delta h \Delta T / T_d^0$ and substituting the values appropriate to polyethylene in xylene of $\Delta h = 280$ J cm^{-3}, $T_d^0 = 114$ °C and σ_e (from melting point data) $= 90$ mJ m^{-2}, gives for a crystallization temperature of 90 °C a value of 21 nm for l^*. The observed interfold length, allowing for the tilt of molecules within a lamella, is 17 nm. Secondly, the strongly negative temperature coefficient growth rate, itself a firm pointer to the involvement of nucleation, is evident. From equation (6.3), and using the same substitution for Δf, the expression $\Delta G^*/kT$ reduces to

$$\frac{32\sigma^2\sigma_e(T_d^0)^2}{(\Delta h)^2 kT_c(\Delta T)^2}$$

of which $(\Delta T)^2$ is the most strongly varying quantity; from equation (6.5) this will govern the overall rate of growth.

6.1.2   Secondary nucleation

It is experimentally evident that secondary, rather than primary, nucleation is more relevant to the determination of lamellar thickness. In this case the relevant nucleus to consider is a rectangular slab, one stem thick, on a substrate of existing crystal. Assuming that the thickness of the substrate is not less than that of the nucleus and that the nucleus is in contact with the substrate everywhere along one side (Fig. 6.1) one has for the excess free enthalpy of formation

$$\Delta G = 2va^2\sigma_e + 2al\sigma - va^2l\Delta f$$

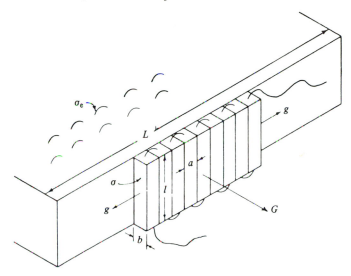

Fig. 6.1. Model for surface nucleation and growth of a chainfolded crystal. (From Hoffman, Davis & Lauritzen, 1976.)

One cannot, however, proceed as before to obtain the dimensions of the critical nucleus by equating $\partial\Delta G/\partial v = \partial\Delta G/\partial l = 0$. This is invalid because it carries the implicit assumption that v is a large number and so may be treated as a continuous variable. This procedure, nevertheless, leads to

$$l^* = \frac{2\sigma_e}{\Delta f}$$

just half the size of the primary nucleus, a well-known result of classical nucleation, and

$$\Delta G^* = \frac{4a\sigma\sigma_e}{\Delta f} \simeq \frac{4a\sigma\sigma_e T_d^0}{\Delta h\Delta T} \tag{6.6}$$

showing a $(\Delta T)^{-1}$ dependence compared to the $(\Delta T)^{-2}$ predicted for primary nucleation. It is easy to see that, in fact, $l^* > 2\sigma_e/\Delta f$ for the polymer case, if this is to be the lamellar thickness. The first of equations (6.6) can be rearranged as

$$\Delta G = a^2v(2\sigma_e - l\Delta f) + 2al\sigma$$

which can never become negative and so allow the crystal to grow if l remains equal to $2\sigma_e/\Delta f$.† An equivalent statement is that a lamella of this thickness would melt at its crystallization temperature because from equation (3.1)

$$T_m = T_m^o\left(1 - \frac{2\sigma_e}{\Delta h l}\right)$$

and setting

$$l = \frac{2\sigma_e}{\Delta f} \simeq \frac{2\sigma_e T_m^o}{\Delta h\, \Delta T}$$

one obtains

$$T_m = T_c$$

This is contrary to observation. It follows that if the thickness, l, of the lamella stays equal to that of the secondary nucleus

$$l > \frac{2\sigma_e}{\Delta f} \tag{6.7}$$

i.e.

$$l = \frac{2\sigma_e}{\Delta f} + \delta l$$

6.1.3   Theory of Lauritzen and Hoffman

The first evaluation of δl was made by Lauritzen & Hoffman (1960) on the assumption that the thickness of a chainfolded strip was invariant after nucleation. Their treatment begins by regarding the folding of a molecular strip on a substrate as the sequential addition of a series of segments and seeks to find the steady state rate of transfer of segments from the melt ($v = 0$) to the first stem in the nucleus ($v = 1$) and thence along the crystallizing strip. A modern version of this theory is as follows.

Consider the laying down of a molecular strip by the following stages: (1) attachment of a segment of length l, width a and thickness b, (2) formation of the first fold and addition of the second segment, etc. At the completion of each of these stages the condition of the growing nucleus is well defined and so is its excess free enthalpy. If contributions other than folding (i.e. chain-ends) to the fold surface free enthalpy σ_e are ignored, the additional amounts after the successive stages are (1) $2bl\sigma - abl\Delta f$,   (2) $2bl\sigma + 2ab\sigma_e - 2abl\Delta f$ and   after v folds

† Classically, l would increase after nucleation.

$2bl\sigma + 2vab\sigma_e - (v+1)\ abl\Delta f$. The free enthalpy differences between successive stages are $2bl\sigma - abl\Delta f$ initially and $2ab\sigma_e - abl\Delta f$ thereafter. The activation energies of attachment and detachment of the successive segments are not, however, well defined. They depend upon what is usually called the apportionment of the free enthalpy of crystallization between forward and backward steps. According to the intermediate stages and configurations through which a segment passes before its complete crystallization, it is conceivable that the maximum activation barrier to be surmounted could occur anywhere between two successive stages. It has usually been taken midway but a general apportionment would be as in Fig 6.2, whereby a fraction ψ of this energy is given to the forward reaction. The activation barrier for the stages $0 \rightarrow 1$ is thus $2bl\sigma - \psi abl\Delta f$ and for the reverse stage $1 \rightarrow 0$, it is $(1 - \psi)\ abl\Delta f$. The difference remains, as it must, $2bl\sigma - abl\Delta f$, the free enthalpy gain between stages. Similarly, for stage $1 \rightarrow 2$ and all subsequent ones the activation energy is $2ab\sigma_e - \psi abl\Delta f$ and, in reverse, $(1 - \psi)\ abl\Delta f$.

One may, therefore, write expressions for the rates between successive stages as:

For $0 \rightarrow 1$

$$A_o = \beta \exp -(2bl\sigma - \psi abl\Delta f)/kT$$

where $\quad \beta = \dfrac{kT}{h} \exp -\dfrac{\Delta\phi}{kT}$ \hfill (6.8)

as in classical theory,

For $1 \rightarrow 2$ and all further stages

$$A = \beta \exp -(2ab\sigma_e - \psi abl\Delta f)/kT \tag{6.9}$$

For all reverse steps

$$B = \beta \exp -(1 - \psi)abl\Delta f/kT \tag{6.10}$$

Notice that the ratios A_o/B and A/B are independent of ψ: only the absolute rate constants depend upon this parameter.

The problem of crystallizing a molecule has thus become the problem of finding a solution to the successive rate equations

$$\begin{aligned} S &= N_o A_o - N_1 B \\ &= N_1 A - N_2 B \\ &= N_2 A - N_3 B, \text{ etc.} \end{aligned} \tag{6.11}$$

where S is the steady-state constant flux of segments, with occupation numbers N_o, N_1, N_2, etc., in successive stages. Adapting the original argument of Hoffman, Davis & Lauritzen (1976), one may obtain this as

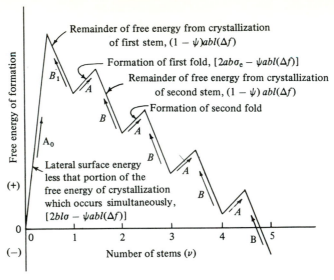

Fig. 6.2. Free energy of formation of a chainfolded surface nucleus. (From Hoffman, Davis & Lauritzen, 1976.)

follows. For $v > 1$, equation (6.11) may be written as $S = N_v A - N_{v+1} B$ i.e.

$$N_{v+1} = \frac{A}{B} N_v - \frac{S}{B}$$

On substituting the equivalent expressions for N_v and then N_{v-1} this becomes

$$N_{v+1} = \left(\frac{A}{B}\right)^2 N_{v-1} - \frac{S}{B}\left(1 + \frac{A}{B}\right)$$

$$= \left(\frac{A}{B}\right)^2 N_{v-1} - S\left(\frac{\frac{A^2}{B^2} - 1}{A - B}\right)$$

and

$$= \left(\frac{A}{B}\right)^3 N_{v-2} - S\left(\frac{\frac{A^3}{B^3} - 1}{A - B}\right)$$

Continuation eventually gives

$$N_{v+1} = \left(\frac{A}{B}\right)^v N_1 - S\left[\frac{\left(\frac{A}{B}\right)^v - 1}{A - B}\right]$$

$$= \left(\frac{A}{B}\right)^v \left(N_1 - \frac{S}{(A-B)}\right) + \frac{S}{(A-B)}$$

Because N_{v+1} must be finite for all v, the coefficient of $(A/B)^v$ has to be zero, giving $N_1 = S/(A-B)$, which, when combined with $S = N_0 A_0 - N_1 B$ from equation (6.11) leads to

$$S = N_0 A_0 \left(1 - \frac{B}{A}\right) \tag{6.12}$$

This answer implies that, of the $N_0 A_0$ chains which start to grow, $N_0 A_0 B/A$ remelt during growth. Explicitly, substituting equations (6.8)–(6.10) in equation (6.12) gives

$$S = S(l) = N_0 \beta \{ \exp(-2bl\sigma - \psi abl\Delta f)/kT \} \times \\ \{1 - \exp -(2ab\sigma_e - abl\Delta f)/kT\} \tag{6.13}$$

The average value of l may be obtained by using S as the weighting factor, i.e.

$$\langle l \rangle_{av} = \int_{2\sigma_e/\Delta f}^{\infty} lS(l)dl / \int_{2\sigma_e/\Delta f}^{\infty} S(l)dl$$

where the limits are taken between the minimum and maximum possible thicknesses and l is taken as being restricted to integral values of the monomer length. The answer is that

$$\langle l \rangle_{av} = \frac{2\sigma_e}{\Delta f} + \left(\frac{kT}{2b\sigma}\right) \frac{2 + (1 - 2\psi)a\Delta f/2\sigma}{\{1 - a\Delta f\psi/2\sigma\}\{1 + a\Delta f(1 - \psi)2\sigma\}} \tag{6.14}$$

For $\psi = 1$ this reduces to

$$\langle l \rangle_{av} = \frac{2\sigma_e}{\Delta f} + \left(\frac{kT}{2b\sigma}\right) \frac{\dfrac{4\sigma}{a} - \Delta f}{\dfrac{2\sigma}{a} - \Delta f} = \frac{2\sigma_e}{\Delta f} + \delta l$$

with the consequence that δl becomes infinite when $\Delta f = 2\sigma/a$, i.e.

$$\frac{\Delta h}{T_d^0}\Delta T \simeq \frac{2\sigma}{a}$$

and

$$\Delta T \simeq \frac{2\sigma T_d^o}{a\,\Delta h} \tag{6.15}$$

For polyethylene in xylene, and taking $2\sigma/a = 40\ \text{J cm}^{-3}$, $\Delta T \sim 55\ \text{K}$, i.e. $T_c \sim 60\ ^\circ\text{C}$. Crystallization at such low temperatures is extremely rapid and difficult to achieve with confidence in practice. There is, however, no experimental evidence for any increase in lamellar thickness from falling crystallization temperatures.† In particular, crystallization of polystyrene from dilute solution, which is sufficiently slow to allow growth at very high supercoolings, shows that the lamellar thickness levels off to a limiting value (Fig. 6.3).

This disparity between theory and practice can be avoided formally by setting $\psi = 0$, which is the reason for introducing ψ specifically into the theory, so that equation (6.14) becomes

$$\langle l \rangle_{av} = \frac{2\sigma_e}{\Delta f} + \left(\frac{kT}{2b\sigma}\right)\frac{\dfrac{4\sigma}{a}+\Delta f}{\dfrac{2\sigma}{a}+\Delta f}$$

which predicts no 'δl catastrophe'.

The physical significance of setting $\psi = 0$ has been suggested to be physical adsorption prior to crystallization, i.e. the molecule would become attached to the face with zero momentum perpendicular to it, before adopting the extended conformation required for crystallization. This, it is proposed, would create the surface energy term $2\sigma bl$ before the compensating free energy of crystallization is released. Certainly the procedure is effective and Fig. 6.3 shows how well data on polystyrene are fitted by making ψ small.

The theory as presented applies only to infinitely long molecules. It is possible, however, to develop it to apply to finite chain lengths. In this case it is found that, for conditions favouring a small number of folds, i.e. short molecules and low supercooling, the lamellar thickness is no longer predicted to be a continuously varying function of crystallization temperature. On the contrary, the excess free enthalpy will clearly take minimum values when chain-ends lie in the lamellar basal surfaces. The expectation is that crystals will contain either fully-extended molecules,

---

† Solution-grown crystals of low molecular mass polyethylene can show a thicker border but this is believed to be due to the crystallization of oligomeric paraffinoid molecules in fully-extended conformation.

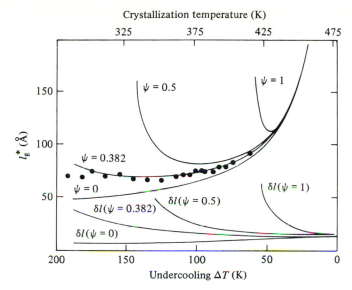

Fig. 6.3. Lamellar thickness of isotactic polystyrene grown from dilute solution and theoretical curves for the values of ψ indicated. (From Hoffman, Davis & Lauritzen, 1976.)

or those folded once into a hairpin conformation, or those folded twice, etc. This is the case for monodisperse oligomeric polyoxyethylenes containing large end groups. Fig. 6.4 shows crystals grown isothermally partly in the fully-extended and then in the once-folded form. Corresponding growth rate curves are shown in Fig. 6.5. This example is probably an optimum; polydispersity would tend to blur the effect as would greater chain length and more accommodating end groups. Thus the once-folded or twice-folded molecules contained in anabaric polyethylene lamellae only tend to have their end groups excluded from the crystal interior. For rapid crystallization of polydisperse samples they are found, in general, to be inside lamellae. With sharper fractions and slower crystallization, however, the end groups become excluded from lamellae. This is very reasonable behaviour and doubtless also reflects the comparatively modest contribution to the total free enthalpy of the system which end groups will make in such thick crystals.

At this point mention should be made of theories of sequential growth due to Sanchez and di Marzio (reviewed in Sanchez, 1974). These were concerned specifically with the effects of incorporating molecular cilia into a crystallizing strip. They produce expressions of the general form of sequential rate processes akin to equation (6.12) but with the first

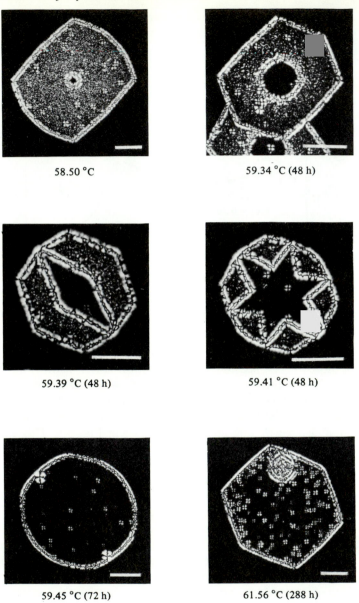

58.50 °C

59.34 °C (48 h)

59.39 °C (48 h)

59.41 °C (48 h)

59.45 °C (72 h)

61.56 °C (288 h)

Fig. 6.4. Habits of polyoxyethylene crystals grown from the melt *in vacuo* at the temperatures shown in the region where increasing times and temperatures cause a transition from once-folded to fully-extended conformations. Crossed polars; bars are 20 μm. (From Kovacs, Gonthier & Straupe, 1975.)

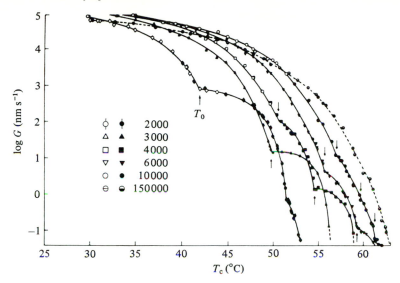

Fig. 6.5. Growth rates as a function of temperature for polyoxyethylene fractions, of the molecular masses marked, with cusps demarcating regions of different molecular conformation. The data marked 6000 refer to the material of Fig. 6.4. (From Kovacs, Gonthier & Straupe, 1975.)

reverse step differentiated from subsequent ones, and with the constants referring to whole segments. One result of general importance is that similar expressions result whether or not cilia are allowed to incorporate, i.e. effectively whether or not there is folding with adjacent re-entry. More specifically, the theory can account for the observed molecular mass and concentration dependence of the kinetics of solution growth. The rate goes through a maximum with increasing molecular mass to $\sim 10^5$, which is explained as a compromise between factors favouring long and short molecules respectively. On the one hand is the increased rate for long molecules because of their greater effective supercooling due to a rise in T_d^0. On the other is the lesser conformational distortion suffered by a smaller molecule next to an interface plus the reduction in σ_e when a chain-end rather than a fold is present at the surface. (This last is the effect revealed by the adoption of a vertical structure for crystals of the shortest polyethylene molecules.)

The concentration dependence comes about because of the possible need to stabilize a growing molecular strip by incorporating new molecules from solution. Whereas this will not be necessary for very long molecules, the shorter the molecule the more it will need additional help to make a sufficiently wide and stable strip. The consequent prediction is that the power α of concentration necessary should increase from zero to values > 1. The experimental results first of all gave $\alpha \sim 0.3$ for an unfractionated polyethylene but later work using narrow fractions found that α increased from 0.2 to 2 as predicted (Fig. 6.6). The incorporation of cilia thus appears to be of considerable consequence in crystallization, at least from solution.

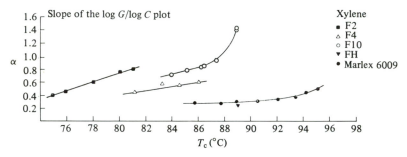

Fig. 6.6. The parameter α for the concentration dependence of growth rate for polyethylene crystals grown from solution in xylene, plotted as a function of crystallization temperature. (From Keller & Pedemonte, 1973.)

6.1.4   Theories with fluctuating fold lengths

The prediction of the Lauritzen–Hoffman theory is of a crystal with ribbons of folded molecules, each of constant thickness, but with variation allowed between different ribbons. The variability arises from a fundamental feature of kinetic theories whereby the thinner the nucleus the faster is the first stage but the slower the second (with the rate, in this case falling to zero at $l = 2\sigma_e/\Delta f$). There is thus a certain variation of rates and corresponding thicknesses around the maximum. The estimated variation from the Lauritzen–Hoffman theory is of order 1 nm. This picture of a crystal derives, however, from mathematical convenience. A more realistic, but mathematically more complex, model is to allow stems in the same nucleus to fluctuate in length.

The first kinetic theory allowing fluctuations was that by Frank & Tosi (1961) who permitted a folded molecular ribbon to change its thickness just once during growth. They were able to show that there is a mean thickness such that if the fold length increases above the mean value then the next fluctuation is more likely to be downwards and vice versa. This is particularly important in giving reassurance that fluctuating fold lengths give a stable mean lamellar thickness. *A priori* it could have been thought that once the fold length began to increase it would continue to do so without limit. Frank and Tosi also point out that the stable fold length calculated for an infinitely thick substrate will be greater than that for a layer nucleating on a lamella of commensurate thickness. The reason is that in the latter case there is the probability of folds protruding above the level of the substrate, at a much enhanced cost in extra free enthalpy because of the additional surface created.

A theory allowing fluctuations in length for every fold stem has been given by Lauritzen & Passaglia (1967) as a particular example of a general solution to sequential chain processes formulated by Lauritzen, di Marzio & Passaglia (1966). This also leads to an expression for lamellar thickness of the form

$$l = \frac{2\sigma_e}{\Delta f} + \delta l$$

or

$$l = \frac{C_1}{\Delta T} + C_2$$

with C_1 and C_2 approximately constant, but it has not yet been possible to avoid the catastrophe of having δl rise to infinity at finite supercooling. Unlike the theory presented in Section 6.1.3, this arises from variations in total flux resulting from steps other than the first. The type of fold surface resulting from successive fluctuations is likely to be a rough one as laid down and, for this reason, the theory invokes a kinetic value of the fold surface free energy $\sigma_e(k)$ which, at high supercoolings, is greater than σ_e for a smooth surface. It is, moreover, possible to allow for smoothing of the surface after a molecule is added, to simulate the effects observed in practice. This is still a relatively minor adjustment, however, and does not cater for the considerable alterations which are observed to occur during growth from the melt.

6.1.5   Kinetic theories of crystallization from the melt

The application of kinetic theories to crystallization from the melt is an empirical rather than a fundamental matter because the formulations presented above are based on the premise that molecular conformations remain essentially as they were laid on the growth surface. When the melt solidifies there is substantial thickening of lamellae during growth behind a thinner growth edge. What has been done, nevertheless, is to carry over the formalism of the existing theories, but to relate it to an actual thickness increased by a factor γ over that of the initial strip. This simple expedient accords with what is known of the facts, but it has, as yet, received no explanation.

Comparison of actual lamellar thicknesses grown from the melt with those predicted on kinetic theories give values for γ often between 2 and 5. It is ~ 2–2.5 for polyethylene under normal conditions, ~ 2.5 for anabaric polyethylene, and ~ 2.6 for PTFE. Alternatively, one may substitute

$$l = \gamma l_g^* \simeq \gamma \frac{2\sigma_e T_m^{\,\circ}}{\Delta h(T_m^{\,\circ} - T_c)}$$

when δl is small in the melting equation given by equation (3.1)

$$T_m = T_m^{\,\circ}\left(1 - \frac{2\sigma_e}{l\Delta h}\right)$$

to give

$$T_m = T_m^{\,\circ}\left(1 - \frac{1}{\gamma}\right) + \frac{T_c}{\gamma}$$

If γ is constant, a plot of T_m against T_c will be linear which, to a fair approximation, it is in practice. PCTFE is a case in point, giving $\gamma = 3.4$. Moreover, the intersection with the line $T_m = T_c$ will occur at $T_m^{\,\circ}$. The values so obtained are near to, but may well differ by a few K from those given by a plot of T_m against $1/l$ even for the same samples.

Kinetic theory continues to be used for melt-crystallized samples in ways analogous to those for solution crystallization. Much of the reason for this has been the considerable difficulty in producing firm facts to show how the behaviour differs in the two cases. Now that the morphology of bulk polymer can be well characterized using the new electron microscopic techniques the situation will alter and, in the author's opinion, is likely to stimulate further theoretical activity in this important area.

6.2   Theory of the growth rate

The kinetic theories of crystallization lead naturally to predictions of the temperature dependence of the growth rate of a crystal. Such data exist to a limited extent for lamellae formed in solution and in considerable quantity for spherulites whose radial advance is one of the most precisely measurable, and measured, morphological quantities. Knowing the rate of nucleation of new layers at the crystal edge, i.e. the total flux already calculated, one needs only to relate this to the manner of completion of the layer to obtain the rate of advance by the molecular thickness. There are two extreme cases which have become known as regimes I and II. Regime I is when completion of a layer is rapid compared to the nucleation rate so that one nucleus will suffice per strip

whereas in regime II the relative rates are reversed and multiple nucleation is expected.

6.2.1   Regime I growth

A layer of thickness b will advance at a speed

$$G = bS_T\frac{n_s}{N}$$

where n_s is the number of sites associated with a particular nucleus and
S_T is the total flux obtained by integration of equation (6.13). Thus

$$S_T = \frac{1}{l_u}\int_{2\sigma_e/\Delta f}^{\infty} S(l)dl \tag{6.16}$$

where the integral gives the sum over all values of l in excess of $2\sigma_e/\Delta f$,
each being an integral multiple of the monomer length l_u. Substitution of
equation (6.13) into equation (6.16) leads to

$$S_T = \frac{N_0\beta}{l_u}P \exp(2ab\sigma_e\psi/kT) \exp\left(-\frac{4b\sigma\sigma_e}{\Delta fkT}\right)$$

where

$$P = \frac{kT}{2b\sigma - ab\Delta f\psi} - \frac{kT}{2b\sigma + (1-\psi)ab\Delta f}$$

The second exponential term, $\exp(-4b\sigma\sigma_e/\Delta fkT)$ is the same as $\exp(-\Delta G^*/kT)$ found classically for secondary nucleation (equation (6.6))
and dominates the temperature dependence for low supercooling. At
lower temperatures, transport becomes more important and, for many
polymers – though not polyethylene – then becomes dominant. For
crystallization from the melt it is usual to set

$$\beta = \left(\frac{kT}{h}\right)J_1 \exp\left\{-\frac{U^*}{R(T-T_\infty)}\right\} \tag{6.17}$$

with the exponential term reflecting the empirical behaviour of the
fluidity of molten polymers; T_∞ is a temperature some 30 K below the
glass temperature and U^* an energy close to 6 kJ mole^{-1}; J_1 has the form
$\exp-(\Delta F^*/RT)$ and refers to all remaining barriers. The corresponding
expression for solution growth is

$$\beta = c'\frac{kT}{h} \exp -\left(\frac{\Delta H^*}{RT} - \frac{\Delta S^*}{R}\right)$$

with c' a function of the polymer concentration and ΔH^* the activation energy of diffusion.

Combining these various expressions one reaches

$$G = G_{o1} \exp -\left(\frac{U^*}{R(T-T_\infty)}\right) \exp -\left(\frac{4b\sigma\sigma_e}{\Delta f k T}\right) \qquad (6.18)$$

with

$$G_{o1} = b\left(\frac{kT}{h}\right) n_s J_1 \exp\left(\frac{2ab\sigma_e \psi}{kT}\right) \qquad (6.19)$$

for the bulk polymer and observing that $P \sim l_u$. We may also set

$$n_s a = L \qquad (6.20)$$

where L is the distance between adjacent nuclei on the growth front; L has been predicted and observed, in polyethylene, to be $\sim 1\ \mu m$.

6.2.2   Regime II growth

When there is multiple nucleation, at a rate

$$i = \frac{S_T}{N_o a} \qquad (6.21)$$

forming centres which then spread laterally at constant height and speed g, the rate of completion of a layer will be proportional to $(ig)^{\frac{1}{2}}$ so that

$$G = b(ig)^{\frac{1}{2}} \qquad (6.22)$$

for regime II conditions. (Use of equation (6.21) gives a comparable expression $G = biL$ for regime I.) One may identify $g = a(A-B)$ in terms of the nucleation process, i.e.

$$g = a\beta\left[\exp\left\{-\frac{2ab\sigma_e}{kT} + \frac{\psi abl\Delta f}{kT}\right\} - \exp\left\{-\frac{(1-\psi)abl\Delta f}{kT}\right\}\right]$$

using equations (6.9) and (6.10). Setting

$$l = \left(\frac{2\sigma_e}{2\Delta f}\right) + \delta l \quad \text{gives } g \simeq a\beta \exp\left\{-\frac{2ab\sigma_e(1-\psi)}{kT}\right\} \qquad (6.23)$$

so that from equation (6.23) and substituting for S_T and β in equation (6.22) gives

$$G \simeq b\beta \exp\left\{(2\psi-1)\frac{ab\sigma_e}{kT}\right\} \exp -\left(\frac{2b\sigma\sigma_e}{\Delta f k T}\right)$$

$$= G_{\text{oII}} \exp -\left\{\frac{U^*}{R(T-T_\infty)}\right\} \exp -\left(\frac{2b\sigma\sigma_e}{\Delta f k T}\right) \tag{6.24}$$

with

$$G_{\text{oII}} = b\frac{kT}{h}J_1 \exp\left\{(2\psi-1)\frac{ab\sigma_e}{kT}\right\} \tag{6.25}$$

and again taking $P \sim l_u$.

It follows from equations (6.19) and (6.24) that $G_{\text{oI}} \gg G_{\text{oII}}$. Moreover, in contrast to growth under regime I where the growing surface is likely to be relatively smooth, that under regime II will be rough because of the multiple nucleation involved. This may lead to observable morphological differences and, indeed, the transition from ridged to planar then S-shaped sheets in polythylene described in Section 4.5.3 may well derive from just such a change. Although the detailed comparison has not substantiated a precise correspondence of a particular lamellar habit with one or other regime, nevertheless, at an optical level, it is quite evident that polyethylene spherulites formed under regime II are finer, as would be expected for multiple nucleation, than those grown in regime I which are mostly immature and coarsely branched.

The reality of the two regimes is demonstrated by analysis of their respective kinetics. In both cases the growth rate has the form

$$G = G_o \exp -\frac{U^*}{R(T-T_\infty)} \exp -\frac{K_g}{T\Delta T} \tag{6.26}$$

but the respective quantities K_g differ by a factor of two.

For regime I

$$K_{\text{gI}} = 4b\sigma\sigma_e \frac{T_m^o}{\Delta h_f k} \tag{6.27}$$

and for regime II

$$K_{\text{gII}} = 2b\sigma\sigma_e \frac{T_m^o}{\Delta h_f k} \tag{6.28}$$

A suitable graph to test this prediction is obtained by rewriting equation (6.26) in the form

$$\log_e G + \frac{U^*}{R(T-T_\infty)} = \log_e G_o - \frac{K_g}{T\Delta T}$$

Plotting the left hand side against $1/T\Delta T$ should, and does, show a change in slope, by the requisite factor of 2, as the crystallization temperature increases above ~ 127 °C for polyethylene (Fig. 6.7).† The slopes of the plot give values of $\sigma\sigma_e$ (from equations (6.27) and (6.28)) which are internally consistent and in very good agreement with other independent estimates of σ_e and σ.

6.2.3   Homogeneous nucleation

One method giving alternative estimates of the surface free enthalpies is the droplet experiment described in Section 4.3. Subdivision of the melt into spheres of a few microns in diameter isolates heterogeneous nuclei and allows the great majority of droplets to crystallize, after homogeneous nucleation, at temperatures much lower than are otherwise obtainable. The thermodynamics of this process are those described at the beginning of this chapter so that the rate of nucleation is given by equation (6.5), or re-expressing the transport term in the same way as for the growth rate theory, i.e.

$$I = I_0 \exp -\frac{U^*}{R(T-T_\infty)} \exp -\frac{32\sigma^2\sigma_e}{(\Delta f)^2 kT} \qquad (6.29)$$

Plots similar to those for the growth data thus lead to values of $\sigma^2\sigma_e$; these will, however, refer to considerably lower temperatures than for other experiments. There is, nevertheless, pleasing agreement between values for σ and σ_e obtained from combinations of nucleation rate, growth rate, melting data (using equation (3.1)) and also geometrical arguments based on the work of chainfolding. Values for polyethylene give $\sigma_e = 93$ mJ m^{-2} and $\sigma = 14$ mJ m^{-2}, for both solution and melt-growth.

6.3   Kinetics of crystallization

So far in this chapter the theoretical treatments have all been based on the knowledge that polymers form lamellar crystals. The development has been, implicitly or explicitly, in terms of molecular chainfolding though it is not thought that variations in the regularity of folding will cause substantial modification in the form of the equations. There is, however, a body of theory, which substantially predates the discovery of chainfolding and is concerned with the development of crystallinity with time. This is associated particularly with von Goler and Sachs, Johnson and Mehl and Avrami (reviewed in Schultz, 1974, pp. 380–90). The theory is based upon an analysis of the growth from centres, nucleated

† The observable crystallization range does not always encompass the transition so clearly. Some polymers such as PCTFE show mostly regime I kinetics while isotactic polystyrene falls into regime II.

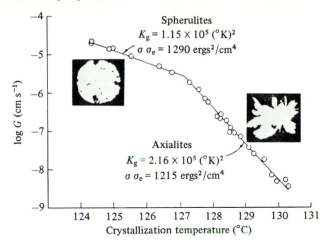

Fig. 6.7. Change in growth regime for melt-crystallized polyethylene for a fraction of molecular mass 30 600 revealed by the altered slope of a plot of log (growth rate) against temperature. (From Hoffman, Davis & Lauritzen, 1976.)

either homogeneously or heterogeneously, and proceeding in a given number of dimensions until impingement, which is considered to stop development. It leads to expressions of the form

$$1 - \chi(t) = \exp -kt^n \tag{6.30}$$

for the development of crystallinity χ with time t where k is a constant at constant temperature. The Avrami exponent n takes on integral, or half-integral, values according to the conditions of growth mentioned above and also whether transport to the crystal (diffusion control) or attachment on the surface (interface control) be the rate-controlling process. For spherulites nucleated sporadically in time growing in three dimensions, and having interface control, the Avrami exponent should be 4. The value should, however, drop to 3 if nucleation were instantaneous.

A great deal of effort, particularly in careful dilatometry, has been expended upon following the progress of polymeric crystallization and analyzing it according to equation (6.30). It has to be stated, however, that the results have not matched the scale of the efforts. What is generally found, and this is hardly surprising given modern understanding of the nature of spherulites, is that n falls during crystallization from initial values of 3–4 and is often non-integral. Fig. 6.8 shows data for linear polyethylene which have this character. It is usual to distinguish the early stages, before the curves deviate markedly from the theoretical

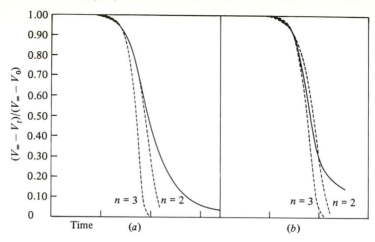

Fig. 6.8. Dilatometric isotherms for the crystallization of fractions of linear polyethylene ($M = 24\,000$ in part (a) and $M = 300\,000$ in part (b)). The dashed lines represent the best fit to the Avrami equation for the values of n shown. (From Mandelkern, 1964.)

isotherms, known as primary crystallization, from the later stages, called secondary crystallization.† By the time the latter starts, spherulites have generally impinged and growth will be either by subsidiary lamellae, in the sense of Chapter 4, or increases in crystallinity associated with isothermal thickening. From a morphological point of view, the fundamental simplicity of the basic theoretical postulates does not seem adequately to reflect the complicated development of spherulites and it is hardly unexpected that there are marked deviations from predicted behaviour.

6.4   Further reading

Much of this chapter is based on the excellent and comprehensive review of kinetic theories by Hoffman, Davis & Lauritzen (1976), pp. 497–614. This and the review of Sanchez (1974) together provide a wide background, linking the various theories and their development; both require only modest mathematics.

A useful summary of the analysis of kinetic data according to the Avrami equation is to be found in Schultz (1974), pp. 380–90.

† This is a different usage from that given in Chapter 2 where secondary growth was defined as interfibrillar growth which is probably to be identified with the solidification of subsidiary lamellae.

7　Crystallization with enhanced chain-extension

As knowledge and appreciation of chainfolded structures has grown, so has the recognition than enhancement of chain-extension might lead to a radical change in both polymeric texture and properties. It has long been believed that this happens to some degree in drawn fibres, providing the basis of their enhanced longitudinal stiffness. In consequence there has been considerable research in the last decade or so into the attainment of high chain-extensions, with concomitant gains in understanding. At the same time what have become known as *ultra-stiff* or *ultra-high-modulus* fibres have been developed in parallel studies. Although much remains to be done, at least the foundations of the underlying morphologies have begun to be established. These concern the later parts of this chapter and of Chapter 9: before that the high-pressure crystallization of polyethylene is discussed. This yields crystals which can be many microns thick but whose formation remained a mystery for ten years. A simple explanation has now been given: that the phenomena result from the intervention in crystallization and annealing processes of a high-pressure disordered phase. This and the comparatively great crystal sizes make this system an informative and seemingly unique model for illuminating many aspects of crystallization behaviour and structure–property relationships, of which there have already been several illustrations in previous chapters.

7.1　Nomenclature

The descriptions *extended-chain* or *chain-extended* have been used in differing contexts in polymers resulting in a certain amount of confusion. To a crystallographer the extended part of an aliphatic chain is its all-trans sequence; in different systems this can vary from a few to at least a few thousand consecutive atoms. To a polymer scientist the terms tend to be contrasted with chainfolded or folded-chain with the implication that folded molecular conformations are (largely) absent. The ultimate *fully-extended-chain* has also been used to refer to examples such as *n*-paraffin crystals where folding is certainly absent. In many cases, however, the location of molecular ends is not known for certain. Following the observation of very thick lamellae in PTFE and polyethylene crystallized at ~5 kbar it has been suggested that it is sufficient to call a crystal extended-chain if it exceeds 200 nm thickness, principally because this is sufficient to depress the melting point by no more than ~1 K. On the other hand, molecules within these layers generally do fold back at the interfaces and the extents of folding and interlamellar connections are important in determining whether such materials are brittle or ductile. To discuss these matters one needs to refer to the folds and it is logically absurd to have to refer to folds in chain-extended samples. Moreover, except in the special low molecular mass polyoxyethylenes studied by Kovacs (see, for example, Kovacs, Gonthier & Straupe, 1975) (Fig. 6.4), crystallization processes have not been shown to yield specific conformations, but rather a variety. The difficulties are compounded if extended-chain is used to describe growth processes in addition to materials. The literature, particularly concerning polyethylene processed at high pressures, is confusing in these respects. In an attempt to avoid these problems the term *anabaric* has been coined, signifying literally 'up-pressure' to refer to the distinct phenomena associated with the crystallization of polyethylene above ~3 kbar. Although the significance is believed to be simply that related to annealing in, or crystallization of, the high-pressure phase, the term is deliberately neutral which would allow modification of this specific hypothesis if it were to become necessary.

7.2   Anabaric crystallization of polyethylene

Wunderlich and co-workers (see Geil, Anderson, Wunderlich & Arakawa, 1964) discovered in 1962 that polyethylene formed very thick lamellae (to ~3 μm) (Fig. 7.1) when crystallized at pressures ~ 5 kbar, dimensions which are commensurate with typical molecular lengths. Partly for this reason and partly because the appearance of fracture surfaces† resembled those of ~ 100 nm thick lamellae grown at 128 °C *in vacuo* from a fraction of mass 12 000 (i.e. close to full extension) they were called, and have become widely referred to as, extended-chain crystals. In fact the platelets generally contain folded molecules, as was soon established by comparison of molecular length and crystal thickness distributions, and also from thicknesses changing with crystallization conditions. More recently, nitration/GPC studies have confirmed this conclusion but shown that fully-extended conformations do exist in particular circumstances, notably when crystals grow very thick, even to many times the lengths of individual molecules.‡

---

† The clarity of fracture surface detail improves as full chain-extension is approached.

‡ Crystal populations to 40 μm thick have been grown from 50 000 mass polyethylene held at 237 °C and 4.8 kbar for 200 h; 20% of lamellae had thicknesses in excess of 10 μm whereas only 1.2% of molecules were longer than this dimension.

Fig. 7.1 Fracture surface replica of linear polyethylene crystallized from the melt at 4.95 kbar.

7.2.1   The high-pressure hexagonal phase

Modern understanding of these phenomena began with a proposal by Bassett & Turner (1972) that they involved the intervention of a then hypothetical new phase in crystallization at sufficiently high pressures. A phase with the predicted properties was subsequently discovered by Bassett, Block & Piermarini (1974). The single X-ray line recorded for this phase (Fig. 7.2) was sufficient to show that the structure had the hexagonal symmetry appropriate to the packing of rods. Coupled with knowledge of the increase in volume over the orthorhombic phase, it also allowed the inference that the all-trans conformation of the chain had been lost. Subsequent work has confirmed these findings. Sharp X-ray reflexions are found only on the equator (the chain direction being called vertical) corresponding to successive orders of rod packing. All

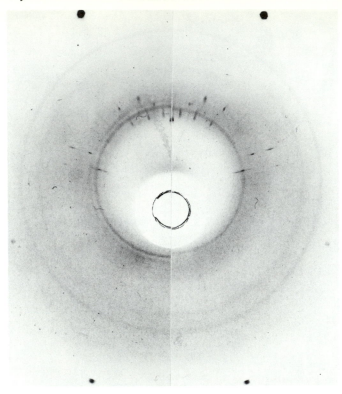

Fig. 7.2 Composite X-ray patterns of polyethylene recorded at high pressure and temperature in the diamond-anvil cell. On the left are the 110 and 200 rings of the orthorhombic structure, on the right the single 100 hexagonal ring to which they give way before the polymer melts. The two large-diameter outer rings are due to the metal gasket enclosing the polymer; other diffraction features are from the diamonds. (From Bassett, 1976.)

non-equatorial reflexions are diffuse, so no ordered repetitions exist along the chain direction. The structure thus falls within the categories of liquid crystal: it is not nematic because of the sharp X-ray lines but would be allotted to the smectic B classification. Internally, the molecules are precisely normal to the layers because when viewed between crossed polars in the diamond-anvil cell the lamellar normal is always an extinction direction.

Hexagonal structures are not uncommon in aliphatic compounds as well as in polyethylene. The best known is the so-called rotator phase of the n-paraffins, which exists close to the melting point in, for example, n-$C_{32} H_{66}$. There is also the structure produced in polyethylene by irradiation of the orthorhombic phase to high doses. All of these may not be unrelated, but certainly the various paraffin rotator phases are to be considered as regular solids. Their specific volumes, v, and entropies, s, are closer to those of the orthorhombic phase than of the melt. For the high-pressure phase the reverse is true with, for example, some 80% of the entropy, and 75% of the volume, of fusion of the orthorhombic phase being required to attain the hexagonal structure at 5 kbar. Such behaviour is common for liquid crystals and imparts a characteristic pattern to the melting endotherm with a relatively small final peak (Fig. 7.4). Moreover, as the various hexagonal phases have lower densities than their orthorhombic precursors the application of pressure makes positive contributions to their free enthalpies. One finds, accordingly, that most hexagonal phases disappear (by 2–3 kbar) as the pressure is increased. This is not true of the high-presssure hexagonal phase whose temperature interval of existence increases with pressure, at least in the range 3–6 kbar. The significance of this can be shown from the Clausius–Clapeyron equation to be that

$$\frac{s_h - s_o}{s_m - s_0} > \frac{v_h - v_o}{v_m - v_0}$$

where the subscripts m, h and o refer to melt, hexagonal and orthorhombic phases respectively. This is in agreement with measurements, confirming that the high-pressure hexagonal phase is one of relatively high entropy. Doubtless the disordered nature of the structure is responsible, with the presence of the gauche bonds possibly being the significant factor.

7.2.2   Crystallization behaviour

The existence of the high-pressure phase allows a straightforward understanding of the patterns of behaviour associated with anabaric crystallization and annealing. The phase diagram has the form of Fig. 7.3 which was compiled from high-pressure DTA melting data of thick, anabarically crystallized lamellae (Fig. 7.4). Melting of samples with thinner lamellae occurs at lower temperatures in accordance with equation (3.1).

The principal feature of the phase diagram is that it leads to the expectation of two regions of behaviour. Well below the triple point (which decreases for higher molecular mass, but is not experimentally well defined) crystallization from the melt will be into the orthorhombic structure and show continuity with atmospheric phenomena. Above the triple point, unless one cools very rapidly into the rhombic-stable region, the hexagonal phase will form with a subsequent transition to orthorhombic as pressure and/or temperature falls. This is what

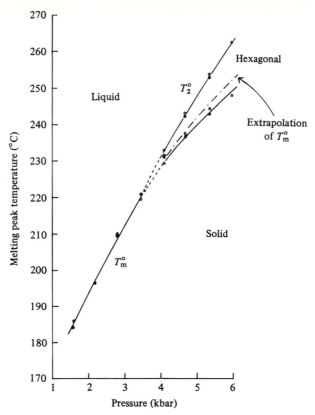

Fig. 7.3. The phase diagram of polyethylene constructed from the observed melting points of anabarically crystallized (at 5 kbar) polyethylene of 5×10^4 molecular mass. (After Bassett & Turner, 1974*b*.)

happens for fractionated or other linear polyethylenes which have no molecules less than $\sim 10^4$ mass. Direct evidence is supplied by high-pressure DTA monitoring of growth. Fig. 7.5*a* shows a single exothermic peak as the orthorhombic phase forms at 2 kbar but two successive exotherms result on cooling at 5 kbar (Fig. 7·5*f*). Discontinuities in volume accompany the DTA peaks.

At intermediate pressures, though seemingly different because of limited resolution, the situation is, in fact, the same. This is strongly suggested by Fig. 7.6 in which the data of Fig. 7.5 are replotted in terms of supercooling. The two peaks evident at 5 kbar coalesce at lower pressures, initially into a clearly broadened composite at ~ 4 kbar (cf. Fig. 7.5*d*) and then persist in unresolved form to below 3 kbar where they

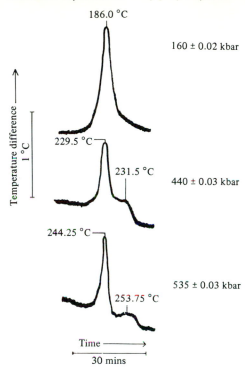

Fig. 7.4. Temperature difference traces produced by anabarically crystallized (at 5 kbar) samples of 5×10^4 molecular mass polyethylene melting at three pressures. (After Bassett & Turner, 1974*b*.)

are joined by an exothermic peak due to direct crystallization of orthorhombic polyethylene (cf. Fig. 7.5*b*). There is then a small region where the two processes both occur at the same pressures until only the orthorhombic phase solidifies from the melt, as in Fig. 7.5*a*. The separate identities of the two processes (i.e orthorhombic growth on one hand and hexagonal into orthorhombic on the other) are indicated by the different supercoolings at which each appears. From Fig. 7.6, orthorhombic polyethylene solidifies from the melt at cooling rates of ~ 1 K min^{-1} some 19 K (at 1 bar) to 15 K (at 3 kbar) less than the maximum observed melting point at the relevant pressure. Hexagonal polyethylene, however, appears at only ~ 12–13 K below. This is a distinction also reflected in the atmospheric melting points: orthorhombic polyethylene which formed via the hexagonal form always melts at a higher temperature than that solidified directly. Fig. 7.7 shows the magnitude of the effect and also how the separate melting populations retain their

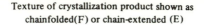

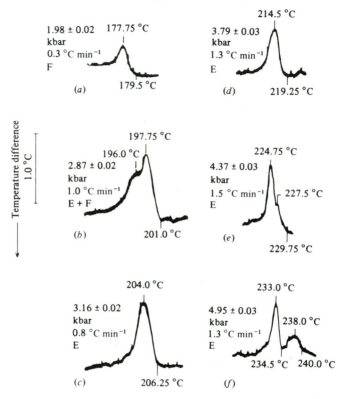

Fig. 7.5. Temperature difference traces recorded during the crystallization of 5×10^4 mass polyethylene at six pressures. (After Bassett & Turner, 1974*b*.)

identities within the narrow (5 K) transitional range of temperatures, at a given pressure, over which isothermal crystallization of the hexagonal form gives way completely to that of the orthorhombic. The difference in melting points relates to the difference in lamellar thickness of the respective populations. In the samples examined there have always been two non-overlapping melting ranges (and, therefore, of lamellar thicknesses) providing a natural division between normal crystallization at low pressures and anabaric behaviour. However, with increasing molecular mass the divide moves to lower pressures and vice versa which alters both melting points and crystal sizes, making it impossible to specify a melting point or lamellar thickness which separates normal and anabaric phenomena in every case. For example, the thickness of lamel-

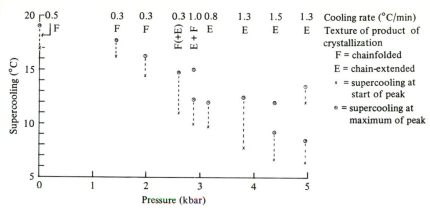

Fig. 7.6. Supercoolings associated with DTA peaks in the crystallization of 5×10^4 mass polyethylene by cooling from the melt under pressure. (From Bassett & Turner, 1974*b*.)

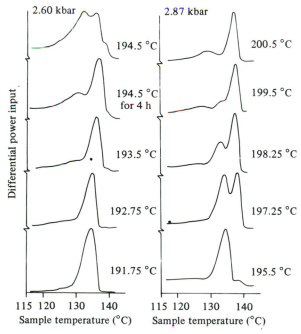

Fig. 7.7. Melting endotherms recorded at 1 bar on 5×10^4 mass polyethylene crystallized in the changeover region of pressures for 2 h (except where stated). (After Bassett & Turner, 1974*a*.)

lae grown anabarically at ~ 3 kbar from 20 000 mass polyethylene is only ~ 80 nm whereas platelets thicker than this can be grown from 30 000 mass fractions *in vacuo*. Any attempted definition of the products of anabaric crystallization in terms of an arbitrary crystal dimension thus cannot correspond to observed behavioural patterns.

One feature which is common to anabarically crystallized polyethylenes and distinguishes them from normal melt-grown polymer is the optical morphology. Viewed between crossed polars the two types of appearance are as in Fig. 7.8; the normal spherulitic texture is quite unlike the coarse spiky microstructure of the anabaric product. The two can even be distinguished when side by side in the same sample (grown at ~ 3 kbar) – termed a mixed product – and can be identified with the respective melting peaks using hot-stage microscopy. The underlying lamellar detail can be revealed either by chlorosulphonation (Fig. 7.9b) or permanganic etching (Fig. 7.9c). One can, moreover, identify in particular instances such as Fig. 5.3 the continued growth of an anabaric lamella as a thin normal platelet to emphasize the fundamental dichotomy at work.

The lamellar sizes produced in anabaric growth extend into the optical range ($\geqslant \sim 1 \ \mu m$) which, with the aid of the diamond-anvil pressure cell (Fig. 7.10) allows considerable further advantages, especially the observation of individual laminae as they form from the melt.

Historically, direct optical investigation came relatively late, casting it mostly in a confirmatory role. Nevertheless, it reveals particularly clearly how the various textures are composed; it has also been able to provide new information not available by other means. The first objects to form at pressures ~ 5 kbar as the isotropic melt is cooled were initially described as cigar-shaped. They are thicker, larger and form more slowly the lower the supercooling and are lamellae viewed approximately edge on, probably as a consequence of preferred molecular orientation in the plane of the specimen imparted by flow during loading. The thinner edge can be seen directly in profile or by the change in birefringence colour in perspectives nearer to the lamellar normals (Fig. 7.11). As previously mentioned the molecular orientation is precisely perpendicular to the platelets. These rarely form regular hexagons but incompletely developed forms with two angles near $120°$ are common. They can be seen in plan and also recognized by a characteristic reduction in birefringence colour towards the ends of the 'cigars' caused by the decreasing optical thickness beyond a crystal's corners. The actual optical appearance is not unlike that reported for smectic layers and, following permanganic etching, can also be viewed with the electron microscope (e.g. Fig. 7.12a).

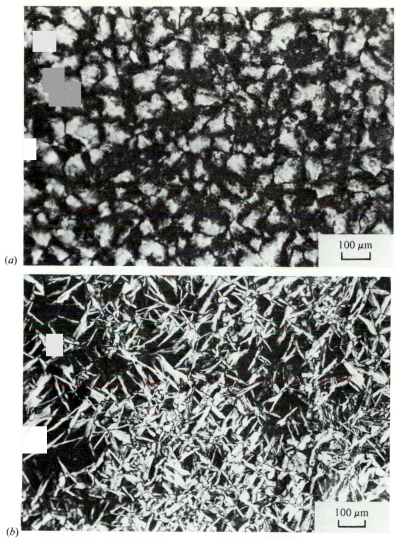

Fig. 7.8. The optical appearance, between crossed polars, of the same linear polyethylene after (*a*) normal crystallization of the orthorhombic phase and (*b*) anabaric growth via the hexagonal structure. (From Bassett & Turner, 1974*a*.)

The hexagonal to orthorhombic transition is recognizable by a sudden change in birefringence colour from a clear overall tone to a confused, often rather muddy, tint. Subsequently, one finds from fracture surface replicas that molecules are no longer always normal to lamellae at room temperature. One presumes, therefore, that the hexagonal to orthorhombic transition involves a tilt of molecules within

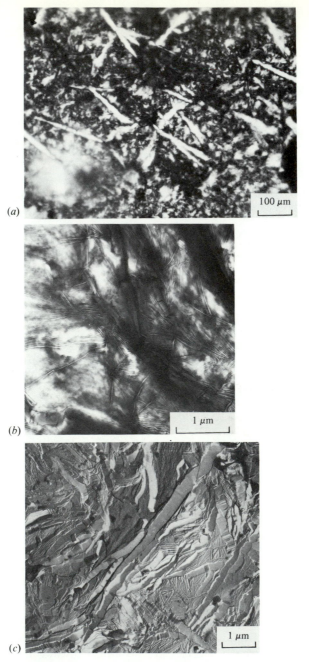

Fig. 7.9. (a) The optical appearance, between crossed polars, of a linear polyethylene ($\bar{M}_m = 4.8 \times 10^4$, $\bar{M}_m/\bar{M}_n = 1.5$) crystallized in the diamond-anvil cell to give a mixed product. (b) The sample of part (a) after chlorosulphonation; electronmicrograph of a stained section. (From Hodge & Bassett, 1977.) (c) A similar sample but containing more anabaric lamellae; etched surface.

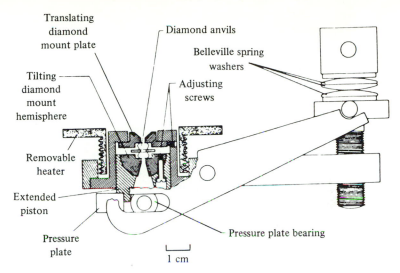

Fig. 7.10. Schematic of the gasketed diamond-anvil cell. (From Barnett, Block & Piermarini, 1973.)

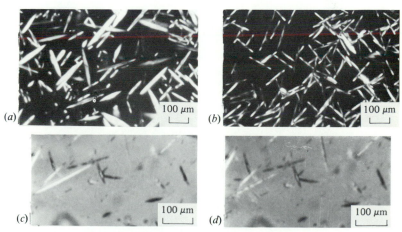

Fig. 7.11. Individual polyethylene lamellae forming from the melt seen between crossed polars in the diamond-anvil cell: (*a*) shows that lamellae are thinner at their edges both from their profile and change of birefringence colour when seen more nearly flat on; (*b*) shows the tendency to develop hexagonal 'corners' revealed by the changing birefringence colour towards lamellar edges; (*c*) and (*d*) show the change in birefringence consequent upon the change of the largest lamella from hexagonal (*c*) to orthorhombic (*d*) phases.

(a)

(b)

Fig. 7.12. Internal boundaries, arrowed, probably resulting from the hexagonal to orthorhombic transition, revealed by permanganic etching (a) parallel to the basal surfaces, and (b) parallel to the chain axis.

lamellae. As any one of the three $\langle 10.0 \rangle$ hexagonal vectors can transform to an orthorhombic $\langle 100 \rangle$ direction, internal boundaries are to be expected. In favourable circumstances these can be watched as they develop; Fig. 7.12 *a* illustrates similar morphology seen in the electron microscope. Occasionally, layers appear to divide in two internally parallel to their length: internal boundaries of this type revealed by their resistance to permanganic etching are seen in Fig. 7.12 *b*. Each half of the original lamella has adopted a different sense of chain tilt. There are, moreover, linear features frequently to be found on the basal surfaces of lamellae (Fig. 7.13) which are reminiscent of lines parallel to *b* running along ridged lamellae as in Fig. 5.2 This suggests, as can be confirmed in particular cases from parallel overgrowths (Figs. 7.14 and 7.9*c*), that lamellae are similarly oblique with {201} sufaces after transformation to the orthorhombic form. It is likely, moreover, that {200} remains the preferred plane of fracture as it does for lamellae grown *in vacuo* whose morphology is exposed predominantly parallel to the spherulite radii. Together with {201} obliquity, this would provide a straightforward explanation of why anabaric lamellae revealed in fracture surfaces tend to show molecules normal to their basal surfaces.

One especially valuable result of observing the optical morphology concerns the changeover from anabaric to normal growth. At 5 kbar X-ray diffraction (using the diamond-anvil cell or equivalent apparatus) shows the presence of the hexagonal phase between the two DTA transitions both in heating and cooling cycles. This is the region where the 'cigar-shaped' objects have crystallized from the melt but not yet transformed. At ~ 3 kbar, however, X-rays have so far failed to detect the hexagonal phase. Nevertheless, optically its role in the formation of anabaric lamellae can still be inferred. Even in a mixed population with only a few per cent of anabaric lamellae, 'cigar-shaped' objects can be seen to form and then to transform with alteration of their birefringence colour (Fig. 7.11). Generally this is on a timescale too rapid for detection with currently available X-ray facilities.

In the region of the changeover anabaric lamellae can be observed forming before the spherulites of normal growth in the same field, in agreement with the supercoolings of Fig. 7.6. The pressures are typically ~ 0.5 kbar below the extrapolated triple point and supercoolings sufficiently low that the orthorhombic phase may only just be capable of crystallizing isothermally (Fig. 7.7*b*). The explanation for the appearance of the hexagonal phase below the triple point appears to be that there is kinetic competition between formation of the (metastable) hexagonal phase and the (stable) orthorhombic form and that the two rates only become equal at conditions sufficiently below the triple point

Fig. 7.13. Surface lines on the fold surfaces of anabaric lamellae presumed to be parallel to *b* and to result from the adoption of obliquity in the hexagonal to orthorhombic transition. Etched surface.

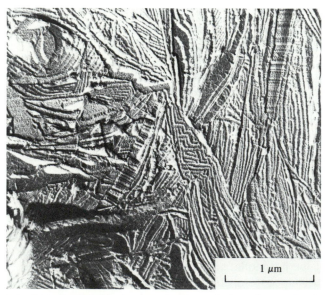

Fig. 7.14. Parallel ridged orientation in anabaric and normal polyethylene lamellae suggestive of a common {201} obliquity.

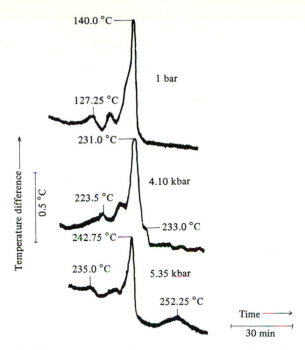

Fig. 7.15. Temperature difference traces produced during the melting of ana-barically crystallized Rigidex 9 polyethylene at three pressures. Note the two low melting peaks due to segregated short molecules. (After Bassett & Turner, 1974*b*.)

to offset the characteristic 3–4 K difference of supercooling between the two processes.

The changeover conditions are, moreover, attainable by rapid increase of pressure from relatively low melt temperatures with its associated adiabatic heating. For example, linear polyethylene with mass average molecular mass $\bar{M}_m \sim 10^6$ commences anabaric growth at 180 °C and 2.3 kbar, values which can be reached by imposing pressure increases of ~ 5 kbar min^{-1} on the atmospheric melt at temperatures as low as 162 °C. This figure rises for shorter molecules, for example to 172 °C for 50 000 mass, but is still within the range of readily accessible circumstances with the implication that at least the natural composites of mixed populations are capable of being produced in conditions not far removed from commercial practice.

When polydisperse polyethylenes are crystallized at high pressures the features already described are retained but are likely to be accompanied by segregation of the shortest molecules into separate populations. When this has happened it is revealed by the presence of lower additional peaks in the melting endotherm, both at high pressures and 1 bar (Fig. 7.15). It is often possible selectively to extract the polymer contri-

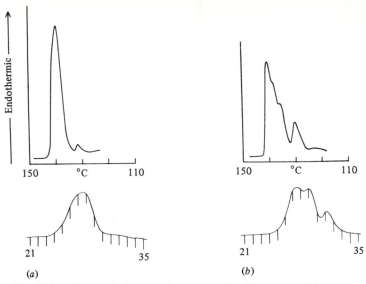

Fig. 7.16. Melting endotherms and corresponding GPC traces of the nitrated polymer for a polyethylene fraction ($\bar{M}_m = 99 \times 10^3$, $\bar{M}_m/\bar{M}_n = 3.2$) crystallized (*a*) at 244.5 °C and 5.5 kbar for 1 h and (*b*) at 226 °C and 4.4 kbar for 1 h. (After Bassett, Khalifa & Olley, 1977.)

Fig. 7.17. Representative morphology of a linear polyethylene fraction crystallized at 2.6 kbar, 60% into the hexagonal phase and 40% into the orthorhombic form. (From Bassett, 1977.)

buting to these extra peaks with controlled exposure to solvent and to measure its molecular mass. For linear polymer at 5 kbar molecules $< \sim 10^4$ mass segregate, a limit which tends to decrease for lower crystallization temperatures and longer times. At lower pressures it increases (by about a factor of two at 3 kbar) often leading, as in Fig. 7.16 to quite complicated melting endotherms. Nitration/GPC (Fig. 7.16) places the lamellar thicknesses of the low peaks usually in the few tens of nm; with permanganic etching the different populations can be identified within the overall morphology (e.g. Fig. 7.17). There is little doubt that these extra peaks are due to polyethylene which has crystallized directly as the orthorhombic phase, though the precise reasons for their number and particular melting points have still to be established. In one sense the cause of the fractionation is that the triple point moves to lower pressures for longer molecules. More fundamentally this probably involves an increases of entropy of the hexagonal phase with molecular mass. At any event the obvious fractionation is of the type where particular species (of short or branched molecules) are unable to crystallize in the time available. Optical observation of developing anabaric textures with the diamond-anvil cell gives the strong impression that there may also be sequential fractionation under isothermal conditions (to parallel the behaviour both from solution and melt at 1 bar, Section 5.2).

7.2.3   Anabaric annealing of polyethylene

Linear polyethylenes can have their average lamellar thicknesses increased continuously from, say, 20 to 200 nm by annealing at successively higher temperatures at 5 kbar. So long as annealing is in the orthorhombic phase the third dimension does not increase beyond 100 nm but passes beyond this figure when the hexagonal phase is formed. One might expect, for example on the local melting plus recrystallization hypothesis, that there would be a sharp increase accompanying the phase change. In practice the transition temperature increases over a range of ~ 10 K with molecular mass and lamellar thickness so that on both counts the actual change should be more gradual, as is observed.

The density and melting point also increase with lamellar thickness but, in polydisperse samples, the melting endotherms become rather complicated. The data in Fig. 7.18 refer to the annealing of a polydisperse homopolymer cold drawn at 1 bar (to give a fibre orientation around the draw direction, which is maintained to the melting point of the hexagonal phase) but otherwise show similar behaviour to melt-crystallized samples. The peak melting points alter as in Fig. 7.19. All peaks other than the highest represent material which is of low molecular mass

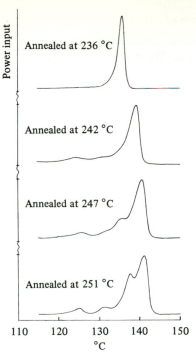

Fig. 7.18. Melting endotherms measured at 1 bar of cold-drawn Rigidex 2 polyethylene after annealing for 15 min at 5.3_5 kbar and the temperatures shown. (From Bassett & Carder, 1973.)

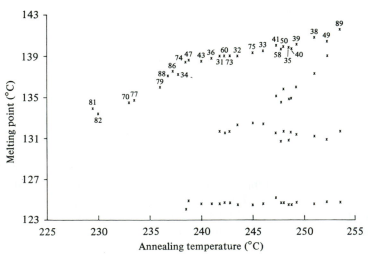

Fig. 7.19. The peak temperatures in Fig. 7.18 plotted as a function of annealing temperature at 5.3_5 kbar. (From Bassett & Carder, 1973.)

and has been molten at the annealing temperature. The two lowest appear near the beginning of the phase change. They are of constant size and melting point and so probably refer to particular short or slightly branched molecular species. Their constituent polymer is unable to adopt the hexagonal structure at 5 kbar and, therefore, melts from the orthorhombic form, flows within the sample and eventually recrystallizes as the orthorhombic form, on cooling. The presence and distribution of all three populations is very important mechanically because they introduce specific modes of brittle failure in tension. The third highest peak has a different character to its two predecessors; it is only resolved well within the hexagonal region and increases both in size and melting point as the annealing temperature rises. The contributing polymer is probably being melted from the least stable components of the hexagonal texture which, on cooling, will allow it the possibility of recrystallizing primarily as the hexagonal form. The high values of lamellar thickness and melting point adopted are evidence that this has occurred.

There is a further feature of annealing which is illustrated best by lightly branched copolymers. Fig. 5.12, referring to ethyl branched polyethylene annealed at 5 kbar, shows a sharp highest peak after treatment at 245 °C. This refers to 15% of the crystallinity and has been shown to contain not more than 3.0 branches per 1000 carbon atoms with, according to electron microscopy of etched specimens, lamellae of thickness to ~ 80 nm. It should be compared with a similar sharp high peak produced at 1 bar in Fig. 5.11. There are two factors which are probably responsible for these well-defined populations. One is that the effective crystallization temperatures and times for the annealing are likely to be greater than for polymer which has become fully molten and had to renucleate to crystallize. The second is that there has evidently been considerable fractionation during annealing. The original segregation anticipated in a melt-crystallized specimen is likely to have been accentuated by the sequence of melting–recrystallization processes which annealing involves, thereby bringing together similar molecules or similar interbranch lengths. The highest peak thus refers to a highly selected component of the sample which is crystalline at the annealing temperature. Lower peaks are of material which probably crystallized less selectively on cooling.

7.2.4   **Summary**

What the existence of the high-pressure phase of polyethylene principally explains is the discontinuity in crystallization behaviour in the region of ~ 3 kbar. This results from the intervention of the hexagonal phase in crystal growth. The high lamellar thickness produced by formation of the hexagonal phase (and subsequently transferred to the orthorhombic) is then made plausible because the entropy of fusion of this phase is only $\sim 25\%$ of the orthorhombic modification. If one can take the same value of surface free enthalpy σ_e in the two cases (which needs justification) then the lamellar thickness of hexagonal crystals should, from equation (6.7), be a factor of about four greater than orthorhombic lamellae formed at the same supercooling. This is the case in the region of the changeover from normal to anabaric growth. In addition, however, there are substantial increases with pressure, ~ 75 nm kbar^{-1}, which raise the average lamellar thickness found at 5 kbar to ~ 500 nm.† These have so far received scant theoretical attention but may well be important in certain other polymers, for example nylons where there are indications, but no more, of appreciable rises in chain extension following high-pressure treatments.

Anabaric phenomena in polyethylene have been considered in detail not primarily for their own sake but because the great associated lamellar thicknesses make them a particularly valuable morphological model system which has given considerable insight into a wide range of polymeric behaviour. Examples include:

(i) the recognition of the tapering growth profile in melt-crystallized lamellae;

(ii) the discovery of the substantial suppression of lamellar thicknesses on crystallizing or annealing in constrained circumstances (small volumes);

(iii) the relationship of annealing to recrystallization;

(iv) fractionation;

---

† Lamellar thicknesses of anabarically grown lamellae depend on molecular length, being greatest for medium molecular masses.

(v) the exclusion of ethyl and longer branches from the polyethylene lattice;

(vi) in mechanical behaviour the unequivocal demonstration that lamellae survive drawing at least eightfold;

(vii) corroboration that chain extension is not of itself a sufficient cause of high stiffness; and

(viii) the identification of certain molecular and textural causes of brittleness.

7.3   Strain-induced crystallization

7.3.1   'Shish kebab' morphologies

The '*shish kebab*', as it is now universally known, is a morphology which was encountered early in the study of polymer systems. It is present, for example, in Fig. 5.21 where it can be seen to be a composite structure of narrow central thread (~ 30 nm in diameter) strung with small platelets. It was mentioned before in connection with interlamellar links (Section 4.4) whose narrow fibrils can become overgrown with platelets in the characteristic fashion when they are allowed to nucleate growth from additional polymer solution. In short the shish kebab is regarded, as it clearly is in the above instance, as nucleation of fibrils followed by epitaxial overgrowth of lamellae sharing a common chain-axis orientation along the fibre. The fibre itself results from strain-induced crystallization, i.e. from distorted molecular conformations in which chains have been brought and maintained parallel for long enough to nucleate crystallites. There are many circumstances where this occurs. In Fig. 5.21 it has probably involved stresses on partly dissolved molecules as lamellae move through the solution in convective flow. Nascent, i.e. as-polymerized, polymer frequently shows this form due, it is believed, to crystallization occurring almost simultaneously with polymerization. Most systematically, however, shish kebabs have been produced by flow-induced crystallization. This is a large subject of great technical importance, being related to processes of moulding and extrusion. Here are considered only the bare essentials involved, especially in so far as they affect the morphology.

Hitherto, the processes of crystallization discussed have referred to unperturbed solutions or melts. It is only necessary to stir these same solutions to permit growth in a higher temperature range which occurs in the form of shish kebabs. For the familiar case of polyethylene in xylene, the upper limit of unperturbed crystallization from dilute solution is ~92 °C, a temperature which is often sufficient to dissolve edges of lamellae formed at lower temperatures, depending on molecular length. General dissolution occurs at ~ 97 °C, with microscopic nuclei, as used in self-seeding, persisting to ~ 105 °C. If, however, a solution is stirred vigorously, fibrous growth will occur to ~ 105 °C; with suitable specific flow fields nucleation has been reported as high as 130 °C. Moreover, precipitation is strongly dependent on molecular length with the longest crystallizing first at a given stirrer speed. Stirrer crystallization, as it is known, has indeed been used as a method of producing reasonable fractions rapidly. Slowly cooling a stirred solution deposits polymer on the stirrer (which is where nucleation occurs) in decreasing order of molecular length. If the stirrer with its overgrowth is then immersed in successive portions of fresh solvent at progressively rising temperatures, a series of increasingly high fractions may be eluted.

The reason for this behaviour is that molecular conformations are being drawn out, the more so for longer molecules. Not only does this tend to bring molecules together parallel, but in so doing it lowers the entropy† of the solution (or melt) and thereby increases the effective supercooling and with it the nucleation and growth rates. Moreover, the extension in the flow field will tend to form a series of nuclei in a row but positively discourage molecular backfolding. One expects, therefore, row nuclei containing reasonably chain-extended molecules on which may deposit, under suitable conditions of low temperatures and modified flow, lamellar overgrowths. So far as has been ascertained this is the case. A schematic structural proposal is shown in Fig. 7.20. The backbone of shish kebabs prepared in this way does not consist of fully-extended molecules, for its melting point is still substantially depressed and, moreover, increases with crystallization temperature, on annealing and with greater stirrer speed, i.e. higher chain-extension. The narrowness of the backbone may be due to the disruption of the initial flow field which it causes but it is so predominant a feature of shish kebabs that a more fundamental origin is likely. Hoffman (1979) has recently proposed that the diameter is limited by cumulative strain arising from what are effectively end-surface regions between successive bundle-like nuclei in the row. It is less clear why the overgrowing lamellae should also appear to be limited in lateral size (< 1 μm), although they can and do form wider units (> 1 μm) when cooling in an incompletely crystallized solution produces extensive overgrowths. Among the suggestions are a decrease in supercooling because of more relaxed molecular conformations or local extensional flow at the lamellar edges.

† It is likely also to lower the enthalpy but not sufficiently to offset the increase in free enthalpy (and consequent rise in dissolution/melting point) due to the decrease in entropy.

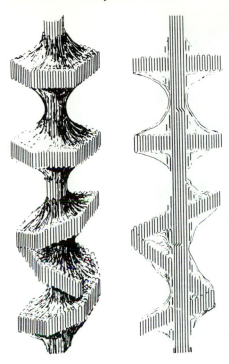

Fig. 7.20. Schematic molecular model of the shish kebab morphology. (From Pennings, 1977.)

These earlier observations have now been supplemented by experiments involving more precise flow fields. It has been found that shear flow does not promote nucleation whereas extensional flow does. The fundamental requirement is that molecules be brought together extended and remain so for long enough to allow nucleation. Shear flow cannot do this, but extensional flow may. Particularly informative have been experiments involving two impinging jets. If polymer is sucked through the jets then, provided these are symmetrically placed, there will be purely extensional flow along the centre line. In suitable circumstances (Fig. 7.21) one can then align molecules with high extension along this axis and, at sufficiently low temperature, initiate crystallization (of shish kebabs).

The suitable circumstances referred to are sufficiently high strain rate and molecular length coupled with a low enough temperature. It can be shown that in a longitudinal, as opposed to a transverse, velocity gradient molecules should become fully extended when the product β of strain rate and relaxation time τ is unity (Fig. 7.22). As the relaxation

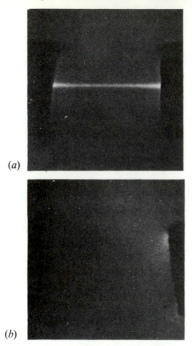

(a)

(b)

Fig. 7.21. Birefringence pattern developed by sucking a 3% solution of Rigidex 2 polyethylene in xylene between opposed jets at 125 °C: (a) polars crossed at 45 °; (b) polars crossed vertically and horizontally. (From Mackley & Keller, 1975.)

time τ depends upon viscosity it increases markedly with molecular mass and the longest molecules become extended first at a given strain rate (Fig. 7.23). It has been predicted that full extension should occur for masses $\geqslant M_c$ at a strain rate $\dot\varepsilon$ such that

$$M_c = (1.1 \times 10^{14}/\dot\varepsilon)^{4/7}$$

for the polyethylene/xylene system. This means that a typical polyethylene with $\bar{M}_m \sim 10^5$ should begin to have its longest molecules extended for a strain rate of $\sim 10^3$ s^{-1}. Two jets with velocity V a distance d apart will have a velocity gradient along the axis of $2V/d = \dot\varepsilon$ so that two flows of 1 ms^{-1} issuing 1 mm apart should achieve a strain rate of $\sim 2 \times 10^{-3}$ s^{-1}. This is approximately the condition of Fig. 7.21 in which the birefringence shows molecular alignment at 125 °C whereas crystallization requires temperatures reduced to ~ 112 °C and less, decreasing with molecular length. The fibrous backbones of such products are believed, on the evidence of their absolute birefringence, to be virtually fully-extended. Even so, they are still overgrown with lamellar platelets,

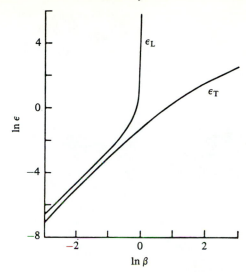

Fig. 7.22. Graph of molecular extension as a function of $\beta = \dot{\varepsilon}\tau$ for longitudinal and transverse velocity gradients. (From Mackley & Keller, 1975.)

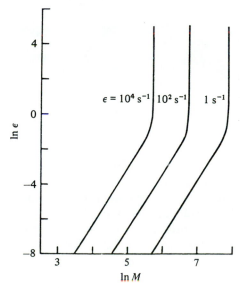

Fig. 7.23. Graph of molecular extension as a function of molecular mass for three strain rates. (From Mackley & Keller, 1975.)

Fig. 7.24. Shish kebab morphology produced by stirring a 5% xylene solution of polyethylene at 510 r.p.m. and 104.5 °C. (From Pennings, 1977.)

which it has not been possible to wash away with fresh solvent at the temperature of formation suggesting that there are molecular connections between filament and platelets. There are fewer platelets for longer molecules, but on the whole the morphology is remarkably insensitive to changes in flow pattern. The nucleation rate is very sensitive, but the result is always rather similar shish kebabs (Fig. 7.24).

A new and related development due to Pennings (1977) is to form fibres in a Couette apparatus, i.e. two concentric cylinders. A seed is attached to the inner cylinder and, on rotation at, for example, 60 r.p.m. and 110 °C using 0.5% polyethylene solution in xylene, fibres form which

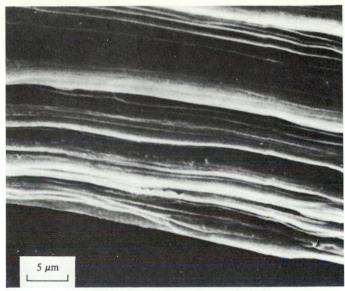

Fig. 7.25. Smooth polyethylene macrofibre grown from a seed crystal close to the rotor surface: xylene solution at 114 °C. (From Pennings, 1977.)

can be reeled up continuously through an aperture in the outer cylinder at rates in excess of 1 m min.$^{-1}$ In these circumstances molecules attached to the rotor are highly deformed and crystallization may proceed to at least 123 °C. The morphology shows shish kebabs for growth at, for example, 107 °C but by 114 °C smooth fibres are observed (Fig. 7.25). Most remarkable, however, are the mechanical properties of the product: a Young's modulus of 110 GPa and high tensile strength at break of 4 GPa. Much attention is now being given to fibre formation in this and related ways which bring the actual properties of polymeric materials much closer to their theoretical limits than hitherto.

7.3.2   Row nucleation

It is now possible to take a more informed look at row nucleation in crystallization from the melt as exemplified by Fig. 2.7a. When the melt is under strain the probability of bringing chains parallel for long enough to nucleate a crystal embryo will increase, as it does for the filaments of shish kebabs. Once a crystallite has formed, by whatever means, its high Young's modulus will accentuate the stress field increasing the probability of a further nucleus along its own axis. The chance of generating a row nucleus in this way will increase with strain and with crystallization temperature (by reducing the free enthalpy driving alternative crystallization). Once a row nucleus has formed it is likely to initiate transverse growth perpendicular to the direction of strain, not least because the nucleated filaments will tend to become load-bearing and so remove stress from residual material. In broad terms this is what happens. Whereas at high extensions the morphology is reminiscent of a collection of parallel shish kebabs, the lamellae become wider ($\gg 1$ μm) as the strain decreases. The central filament may by then either have disappeared (as in natural rubber, presumably because it has become thermodynamically unstable) or simply be obscured. In either event it is frequently the case that melt-crystallized samples appear to be collections of lamellae lying across the direction of stress and strain.

7.4   Gelation

In certain circumstances vigorous stirring of a polymer solution can induce gelation. A gel is a connected network extending through a sample which removes fluidity. A 1% solution of high molecular mass ($> 10^6$) polyethylene in xylene can be made to gel using a Couette viscometer, which provides a shearing flow, and then cooling to 90 °C. The flow appears to have stressed entangled molecules leading, it is suggested, to nucleation occurring at the entanglements. On cooling, the gel has been found to consist of fibrillar crystals.

More familiar circumstances of gelation are those of very high supercooling when network formation occurs before chainfolded crystallization. The situation is particularly well defined for isotactic polystyrene where at sufficient concentration (~ 3–5%) gelation occurs in a temperature range below that of chainfolded crystallization (Fig. 7.26). The network produced is believed to be of fringed-micellar type, connecting bundle nuclei and, it has been suggested, may involve specific stereochemical sequences along the chain. The low temperature of formation is likely to be necessary to allow small crystals to form from such relatively small sequences and also to accommodate the high surface free energy of a bundle nucleus. The gel shows a distinct X-ray diffraction pattern, different from that of the usual isotactic polystyrene structure; it also dissolves or melts considerably below chainfolded lamellae of the same polymer (Fig. 7.27). On heating, the gel suddenly disappears (at ~ 60 °C) and fluidity is restored whereas ~ 120 °C must be reached before the chainfolded lamellae dissolve. When the gel is dried it melts at ~ 120 °C, but the corresponding lamellae only do so at 220–230 °C. There are few, if any, other observable morphological features of gels but they on the one hand and shish kebabs on the other illustrate how the dominant mode of chainfolded crystallization does give way to important rival forms in appropriate circumstances.

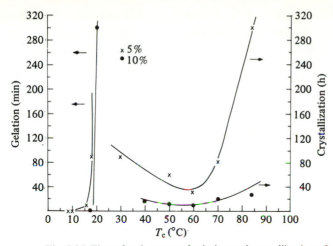

Fig. 7.26. Times for the onset of gelation and crystallization of unfractionated isotactic polystyrene at different temperatures in decalin. (From Girolamo *et al.*, 1976.)

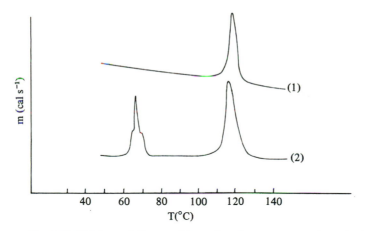

Fig. 7.27. Melting endotherms of isotactic polystyrene suspensions obtained from a 5% solution: (1) turbid suspension formed at 50 °C and then measured; (2) as curve (1) but with cooling to 0 °C, forming a gel, before measurement. (From Girolamo *et al.*, 1976.)

7.5   Further reading

Understanding of the crystallization of polyethylene at high pressures changed abruptly with the prediction and discovery of the anabaric phase. Those who might wish to relate the situations before and after these events will find a review in Bassett (1976).

The paper by Mackley & Keller (1975) should be consulted by those interested in flow-induced crystallization. It reports elegant experiments conducted in purely extensional flow and presents a clear picture of the underlying principles. This route to high-modulus and high-strength fibres has been pioneered by Pennings whose activities are summarized in Pennings (1977).

8 Chemical consequences

Certain chemical consequences are imposed on semi-crystalline polymers by their textural organization. Probably the most important is that the interlamellar – traditionally the 'amorphous' – regions provide diffusion channels through which suitable reagents can more easily penetrate. In consequence many chemical reactions occur preferentially at or near fold surfaces. Moreover, a polymer molecule may, in principle, be inherently more reactive at chain-folds than elsewhere, possibly for steric reasons concerning packing geometry and accessibility to a reactant, or by the introduction of special bond sequences in folding: in polyethylene, for example, gauche bonds are present in the folds but not in fold stems. Furthermore, end groups may be concentrated at or between fold surfaces which, as they are often chemically quite different from the basic monomeric unit, may bring specific and localized chemical reactions to these regions.

8.1 Crystal defects

The total texture of a crystalline polymer, however, goes beyond the factors just listed and includes crystalline defects such as dislocations. In many solids these can be identified and located by etching. There has been a number of elegant studies of dislocation movement carried out in this way on materials such as lithium fluoride. If such effects also occur in polymers, as would be expected, they have still to be observed. Etching various polyolefines has exposed giant screw dislocations at the centre of growth pyramids, but, to a resolution of ~5 nm, no associated etch pits.

There is, nevertheless, one polymeric system in which the chemical specificity of crystal defects has been demonstrated: the solid-state polymerization of trioxane to polyoxymethylene. Solid-state polymerization can convert suitable monomer crystals directly into their corresponding polymer using a variety of initiating techniques, involving chemical, radiation or thermal treatments. There is a great deal of interest in such systems as the polydiacetylenes in which a macroscopic monomer single crystal can transform completely to a single crystal of fully-extended-chain polymer. They may, for example, prove to be capable of showing semiconducting electrical properties, though this is a

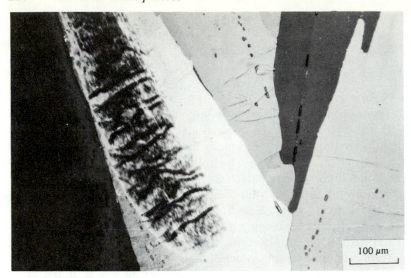

Fig. 8.1. Decoration of transverse, mainly {00.1} subgrain boundaries in triox-ane by polyoxymethylene formed in the solid state. The right-hand grain shows the boundaries just picked out (as dark lines) while that on the left shows substantial development. (From Bassett, 1968*b*.)

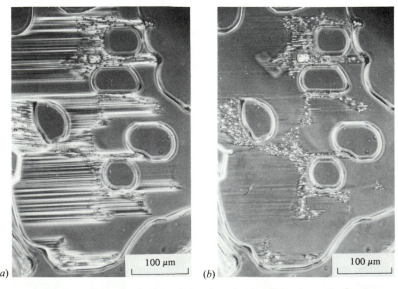

(*a*)   (*b*)

Fig. 8.2. Fibres of polyoxymethylene growing parallel to the *c* axis of a trioxane grain. The orientation is confirmed by this view in phase contrast plus one polar which is parallel to *c* in (*a*) but perpendicular in (*b*). (From Bassett, 1968*b*.)

matter beyond the compass of this work. Trioxane was one of the earliest materials to be studied because of its *topotactic* polymerization, i.e. where the crystallographies of original and product are related. Unfortunately, a single crystal of trioxane yields polycrystalline polyoxymethylene, except at very low yields, with the polymer in twinned orientations; nor does the reaction go to completion. However, following (electron) irradiation and warming, the conversion can be seen to start at {00.1} subgrain boundaries within a monomer crystal (Fig. 8.1); these boundaries may be regarded as collections of (edge) dislocations. Subsequently, reaction proceeds by the development of needles of the polymer parallel to the *c* axis of the monomer (Fig. 8.2). They do so, however, only in the same sense in each crystal: antiparallel to the internal electric field in these polar materials.

8.2   Reactions at fold surfaces

An elegant and convincing demonstration that interlamellar regions provide accessible pathways inside polymer samples is given by the absorption of gas. Fig. 8.3 shows that, for the same polyethylene crystallized from the melt to various levels of crystallinity, the mass of gas absorbed in equilibrium is directly proportional to the 'amorphous'

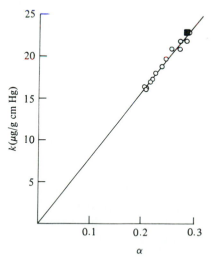

Fig. 8.3. The solubility coefficient k (in μg of gas absorbed per g of specimen) at 80 °C and a gas pressure of 0.2 bar of cyclopropane in melt-crystallized polyethylene as a function of the disordered fraction α. (From McCrum & Lowell, 1967.)

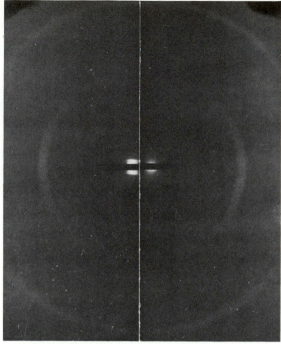

Fig. 8.4. The low-angle and the first two wide-angle reflections for the same sample of poly(4–methylpentene–1) after annealing for 30 min in air (on the left) and in nitrogen (on the right). Note the enhancement of the low-angle reflections produced by oxidation in the left half. (From Bassett, 1964*a*.)

content. In such a highly lamellar material, this part of the sample is associated with the fold surfaces of the lamellae (Section 3.3). Indeed, similar experiments on aggregates of solution-grown lamellae before and after annealing show similar behaviour to that of Fig. 5.18, as the nature of the fold surface changes and eventually becomes like that of a melt-crystallized sample.

With gases and certain liquids able to penetrate between lamellae it would be expected that, if reaction with the polymer is possible, then it will tend to be localized in the fold-surface regions. This has been established unequivocally by low-angle X-ray diffraction. Fig. 8.4 compares the low-angle X-ray patterns of a mat of solution-grown crystals of poly (4–methylpentene–1) after heating in air at 170 °C for 30 min with identical treatment, but in nitrogen. This polymer is very readily oxidized and needs to be stored in the dark *in vacuo* if it is not to be rapidly degraded. In Fig. 8.4 the polymer has a much more intense

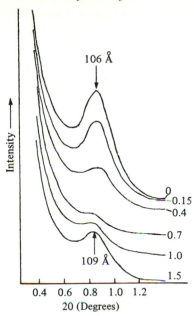

Fig. 8.5. X-ray diffraction traces after bromination of solution-grown polyethylene crystals. The numbers on the right refer to the average number of bromine atoms per fold. Note the initial reduction followed by an increase of intensity. (From Harrison & Baer, 1971.)

low-angle reflection after oxidation than otherwise. This indicates that the oxygen atoms have added to the polymer at the fold-surface regions. As they do this they counteract the initial deficit of electron density associated with the regions through which the gas has penetrated so that the intensity of diffraction goes through a temporary minimum corresponding to near-uniform electron density, before increasing with further reaction. Similar work has also shown that chlorination of poly(4–methylpentene–1) and bromination of polyethylene (Fig. 8.5) occur preferentially in the region of fold surfaces. In the latter case it has been suggested, from infra-red absorption measurements, that there has been selective reaction with gauche–gauche bond sequences in the fold. That is to say reaction is specific to the chain-folds and not merely a matter of localized access of reactant.

8.2.1   Preparation of copolymers

This type of chemical reaction makes it possible to prepare a special type of copolymer, or probably more correctly, substituted polymer with side-groups attached to the main chain at integral separations of the fold-stem length. For example poly(4–methylpentene–1) crystals have been swollen with carbon tetrachloride, then irradiated with electrons which grafts chlorine residues on the fold surfaces of the original polymer. Studies have been made of polyethylene after bromination and after phosphorylation (substitution of hydrogen by the group

$$—P—(OCH_3)_2).$$
$$\overset{\|}{O}$$

The latter is a large substituent so that, although the substituted polymer can melt and recrystallize, once there is approximately one added group per fold, the long period and crystal thickness remain invariant, at slightly above that of the orginal material, with crystallization and annealing treatments. It appears that the bulky side groups can only be accommodated in approximately their original positions. During bromination reaction has been followed in considerable detail with the results shown in Fig. 8.6. The heat of fusion does not change as the bromine content is increased to beyond two atoms per fold, and the long period only slightly. The ability to thicken on annealing is lost progressively and has disappeared by the two bromines per fold level. After melting and recrystallization the heat of fusion has fallen substantially, the more so with greater reaction. The interpretation is essentially as before: that bromine atoms attach to folds but that any repositioning by recrystallization or lamellar thickening increases the free enthalpy of the system – sufficiently to suppress the latter process when there are two atoms added per fold.

8.2.2   Degradation of polyethylene by nitric acid

Treatment of polyethylene with fuming nitric acid is a widely used morphological technique. It has played some part as an etchant for microscopy, albeit a very severe one, but more generally its use is related to its ability to remove fold surfaces and interlamellar regions of the polymer selectively. The acid penetrates into these areas of a sample, digesting material with which it comes in contact. (By comparison, Fig. 7.12 shows that the permanganic etchant makes much smaller inroads between lamellae, which is the basis of its greater utility for microscopy.) It was at one time thought that there was an initial, more rapid, reaction removing folds followed by a slower erosion of the interior, but more detailed measurements have not supported this simple division. In point of fact, polymer is removed most rapidly from side surfaces, i.e. by ablation. This follows because the thickness of lamellar fragments

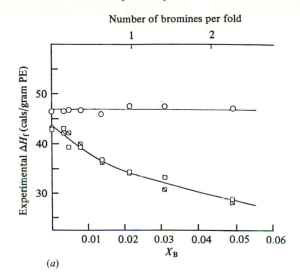

(*a*)

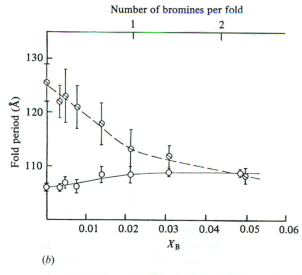

(*b*)

Fig. 8.6. (*a*) Heat of fusion as a function of bromine content X_B for as-prepared polyethylene single crystals (upper line) and the same samples after recrystallization from the melt. (*b*) X-ray long periods before (lower line) and after annealing at 113 °C for 30 min, as a function of bromine content X_B for solution-crystallized polyethylene lamellae. (From Harrison & Baer, 1971.)

decreases much less in proportion than the amount of material lost from a sample. When attempts are made to make weight losses reproducible it becomes clear that the degradation proceeds with considerable subtlety. For example not only does the strength of the acid (in the range ~90 to 100%) influence the rate of attack, it also affects the selectivity of removal of different populations. In samples such as those of Fig. 7.16 and 7.17 the relative proportions of GPC peaks after nitration vary sensitively with the time of reaction and strength of acid. Removal is generally greater for thinner lamellae and lower molecular mass populations.

Fig. 7.16 is one example of probably the major use of nitric acid degradation: the estimation of lamellar thickness or, in the terminology of the two-phase model, the height of the crystalline core within lamellae. (Some qualification is needed because of the diminution of measured lengths with more severe treatments.) Then, for Fig. 3.22, one may proceed to infer something of the nature of the interfacial regions, by comparison of the lengths of double and single molecular traverses of a lamella. With very thick (anabaric) lamellae one may also reveal information on molecular conformations which existed in the solid polymer. If chain-ends are excluded from lamellae, only integral traverses will remain after nitration. Conversely, chain-ends included within lamellar interiors will lead to shorter lengths. In the former case the distribution of stem lengths will, to the approximation that there are no multiple traverses, be the same as that of crystal thicknesses, otherwise residual stems will move to lesser lengths, especially for shorter molecules where the influence of chain-ends is greater. In general, the latter situation occurs following anabaric crystallization (Fig. 8.7) although lengths less than ~60 nm are not observed, i.e. 'walking-stick' conformations are absent. However, with narrower fractions (e.g. a polydispersity of 1.3) good agreement between crystal and stem-length distributions has been found showing that chain-ends have become excluded, probably because like molecules have had a greater opportunity to crystallize together in these circumstances.

For atmospheric crystallization, at least from solution, the evidence is in favour of exclusion of the majority of chain-ends. This has been shown by diffusing ozone, a more delicate alternative to nitric acid, into mats of solution-grown polyethylene lamellae. In the early stages of reaction infra-red examination showed that 90% of the terminal double bonds of commercial linear polyethylene chains were rapidly attacked and lay in planes as would be expected of cilia protruding from fold surfaces (cf. Section 4.4).

Treatment with fuming nitric acid rapidly converts melt-crystallized

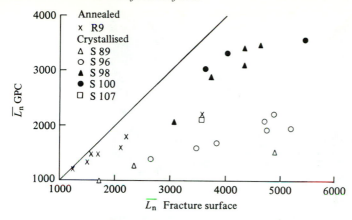

Fig. 8.7. Comparison of number-average lengths measured by nitration/GPC (ordinate) and on fracture surfaces (abscissa) for various anabaric polyethylenes. The numbers attached to the crystallized samples increase with molecular mass. The greater departures from the 1:1 line for shorter molecules reflect their higher proportion of short residual stems left by chain-ends which were included within lamellae. (From Bassett, Khalifa & Olley, 1977.)

and other polyethylene samples into brittle, friable solids which can easily be powdered in the fingers. At an early stage it is possible to recover lamellar fragments which, historically, was important evidence in favour of lamellae existing within melt-crystallized polymer. There can be little or no guarantee that these degraded lamellae (whose folds will probably have been replaced by —COOH groups) will have preserved features pertaining to untreated samples. The lateral habits, especially, can be expected to have largely disappeared. Nevertheless, it has been found that the near {201} orientation of the basal surfaces remains for polyethylenes crystallized at higher temperatures. The average inclination falls to about half this for samples consisting of banded spherulites, which is consistent with S-shaped lamellae as sketched in Fig. 4.31.

The production of brittleness in this way sheds some light on one of the most important weaknesses of polymers: their vulnerability to *environmental stress-cracking*, i.e. failure in particular circumstances, usually involving contact with organic chemicals, which would not be expected in the absence of the particular environment. In a very general sense any material which is not brittle possesses some mechanism of blunting adventitious cracks. In crystalline polymers, crack-blunting is intimately connected with intercrystalline regions. For example, Fig. 8.8 shows how a crack is bridged by fibrils which can still transmit stress.

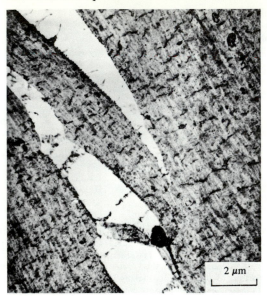

Fig. 8.8 Cracks and bridging fibrils in a detachment replica of drawn, originally spherulitic, film of polyethylene. (From Frank, 1964.)

Were a crack to attempt to propagate between lamellae it would be expected to be similarly bridged by tie molecules or interlamellar links (Section 4.4). Removal of surface regions eradicates these and so allows cracks to propagate unhindered as evidently happens after degradation with nitric acid.

When chemicals are brought into contact with a stressed crystalline polymer – which need only be a moulded unit with internal but no external stresses – its lamellar morphology is likely to bring two consequences. Firstly, it may admit the chemicals to the crucial interfacial regions where they may, over a prolonged period, sufficiently plasticize the local volume that the stress can no longer be withstood. Secondly, the amount of interlamellar linking depends on the ratio of molecular length to crystal thickness which needs to be sufficiently large to maintain ductility. (For example, anabarically crystallized polyethylenes with average lamellar thicknesses of ~ 0.5 μm are brittle until the molecular length exceeds ~ 10 μm.) Lower molecular mass material will thus be more vulnerable to cracking at a given thickness and the tendency for this to crystallize together from the last remaining pockets of melt will concentrate the material most at risk in particular localities. One would expect, therefore, that performance should be ameliorated by the removal of lower molecular mass polymer so as to maintain a high ratio of molecular length to crystal thickness. Although stress-cracking is a complicated and diverse subject involving many more aspects than those mentioned, the precautions proposed are in accord with technological practice.

8.3   Radiation effects

8.3.1.   Cross-linking of polyethylene

High-energy (ionizing) radiations such as electrons, neutrons, X-rays or γ-rays, etc. have profound effects on carbon polymers principally because they alter their molecular constitution. The two major consequences are chain scission, as occurs extensively in polyoxymethylene (Fig. 3.4), and cross-linking which predominates in polyethylene. More generally, and in such crystalline polymers as poly(4–methylpentene–1) and polypropylene, both processes occur to considerable extent. Of particular interest is that the cross-linking of polyethylene is practically confined to the interfacial regions of lamellae, although the radiation interacts with the polymer throughout its volume, with major changes depending upon the details of lamellar packing.

A cross-link, rather like a chain-fold, is a feature which is detected indirectly by its effects rather than by its intrinsic properties. The consequences of cross-linking are twofold. Firstly, each cross-link results in the evolution of a molecule of hydrogen. Secondly, it may lead to gelation by interconnecting previously separate molecules. A gel is effectively an infinite network and requires for its formation one tertiary carbon atom (0.5 cross-links) per mass-average molecule, according to simple theory. The interesting discovery, due to Salovey & Keller, (1961*a*, *b*), is that, whereas the evolution of hydrogen is constant (to within $\sim 15\%$) for linear polyethylene, with the implicaton that roughly the same number of cross-links form in all cases (~ 2 per 100 eV of absorbed radiation), the gel content (i.e. insoluble portion) varies widely with the state of aggregation. At a suitably chosen absorbed dose of radiation, for example 20 Mrad, the gel content can vary between 0 and $\sim 80\%$ for the same polymer (Fig. 8.9). Melt-crystallized samples give high values while solution-crystallized specimens produce results throughout this range depending upon the manner of their preparation.

The clear implication is that, while roughly the same number of cross-links is produced in every case, the proportion which is effective in producing gelation alters. Those which are ineffective must be intramolecular, joining parts of the same molecule, rather than intermolecular

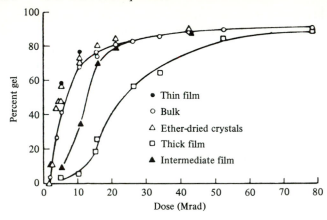

Fig. 8.9. Gel formation produced in polyethylene in various states of aggregation by irradiation. (From Salovey & Bassett, 1964.)

connections. Moreover, a variety of evidence indicates that the site of cross-linking is at or near fold surfaces. In the first instance this was shown by controlled dispersion. For example, polyethylene axialites grown from 1% solution in xylene at 85 °C gave $\sim 60\%$ gel after receiving 20 Mrad irradiation. This figure fell to $\sim 2\%$ if the axialites were previously split into lamellae using ultrasonics, but could be recovered (showing the absence of significant molecular degradation) by recrystallization of the dispersed material with the original conditions prior to irradiation. This shows unambiguously that it is the degree of contact at fold surfaces which is important. Cross-linking in separated lamellae must be almost entirely intramolecular, with intermolecular connections forming only when neighbouring lamellae are in good contact.

This conclusion is reinforced by investigation of different preparative conditions affecting the degree of interlamellar contact. If monolayers are filtered from suspension rapidly with high suction the resulting mat would be expected to have more interlamellar contact and to produce more gel than when the same lamellae sediment slowly: such is the case. Moreover, a sample with poor initial lamellar contact giving low gel formation can have this raised progressively to melt-crystallized values by annealing (Fig. 8.10). Final proof is given by the finding that removal of the surface regions of highly irradiated mats with ozone restores solubility. The cross-linking has disappeared and must, indeed, have been confined to the interfacial areas.

Such behaviour is not restricted to polymers: it also occurs for the dicarboxylic acids of the 'crystal core' left following treatment of chain-folded lamellae with ozone. The probable cause is that irradiation

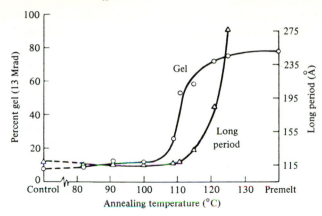

Fig. 8.10. The associated increases of gel content and long period for a particular (intermediate) film of solution-grown polyethylene on irradiation to 13 Mrad absorbed dose. (From Salovey & Bassett, 1964.)

produces free radicals —CH_2—$\dot{C}H$—CH_2— in the chain which must combine in pairs to form a cross-link. They evidently do not do so in the crystal interior to any measurable extent but presumably react when chains are brought closer than the interstem distance, either in chain-folds of the polymer or, more speculatively, at the ends of the dicarboxylic acids where vibrations could also contribute to a reduced separation. Whatever the precise mechanism, it undoubtedly leads to a specific location of the reaction. This in turn is able to explain further elements of behaviour in terms of established features of the morphology.

Adjacent re-entry, for example, is implied by the ineffectiveness of intralamellar cross-linking between adjacent stems, which must also be intramolecular if it is not to lead to gelation. In such conditions, which are believed to be closely approached in freeze-dried preparations of monolayers, it is found that thinner lamellae produce greater gel, probably because of their greater proportion of interfacial regions. Radiation cross-linking is, therefore, a chemical reaction which is specific to the fold surface regions of polymers, but depends on their proportion and their detailed morphology.

8.3.2   Electron microscopy

Radiation damage, be it scission or cross-linking, rapidly causes crystalline polymers to lose their crystallinity under examination in an electron microscope. No way is known of avoiding this difficulty. Even cooling samples to liquid helium temperatures inside the microscope only prolongs their lifetime by a factor of two or three over that for normal conditions. Higher accelerating voltages (e.g. 1 MV) give a similar modest improvement but this is usually offset by the loss of efficiency of conventional photographic emulsions at these potentials, leading to little useful advantage. Normally, photographic plates are very efficient and leave little scope for gains in recording signals using image intensifiers, although these may allow greater time for manipulation. In sum, one must needs rely on rapid working at low beam intensities to examine specimens while they still diffract. For this the recent availability of scanning transmission instruments can be particularly helpful in minimizing damage.

The common result of examining solution-grown polymer lamellae with the electron microscope is to leave a featureless pseudomorph of highly cross-linked polymer. (Scission, as in the case of polyoxymethylene, is more rare.) Particularly in the case of polyethylene, however, a lamella will pass through a number of reasonably well-defined stages as irradiation proceeds. The first may be changes in Bragg contours (diffraction contrast) due to buckling, etc., in the beam, with little or no changes in the diffraction pattern. Eventually spots do broaden, and in polyethylene a hexagonal structure may be formed, especially at higher temperatures. It is not known whether this is a high-volume phase like the rotator phase of the *n*-paraffins or a high entropy one similar to the disordered high-pressure phase. Increasing disordering makes the diffraction pattern disappear, first the interchain packing reflections usually observed and lastly the intrachain spacing. A sequence of diffraction patterns for a polyethylene monolayer is shown in Fig. 8.11.

With lamellae in contact with carbon substrates all the above changes occur at constant area but this is not the case for thin films or other unconstrained specimens. Fig. 8.12 shows the lateral expansion on irradiation of polyethylene monolayers mounted on a film of collodion. This example, recorded at very low temperature, has a lower expansion than normal but demonstrates very clearly that the dimensional change is an inherent property and not dependent on, for example, melting in the electron microscope. Typical figures for the dimensional changes of polyethylene monolayers in normal microscope operating conditions are $+23\%$ in all lateral (radial) directions and a reduction in total volume of $\sim 5\%$ implying a reduction in crystal thickness of $\sim 40\%$, assuming the density is constant which is almost true. A reduction in thickness is inevitable starting from the extended all-trans stem configurations which can only shorten when damaged.

These observations are very relevant to the appearance of thin spherulitic films. Consider an unbanded polyethylene 'spherulite' grown in a thin film, for which *b* is radial but in which *a* and *c* axes are randomly

75 kV

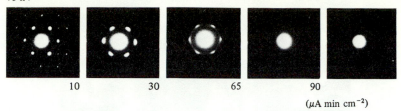

(μA min cm⁻²)

Fig. 8.11. The disappearance of crystallinity for a polyethylene single crystal signalled by its fading diffraction pattern which, in its later stages, exhibits hexagonal symmetry. (From Grubb, 1974.)

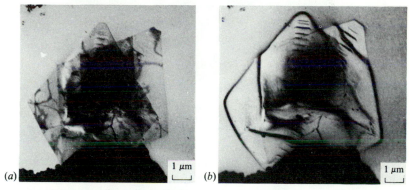

Fig. 8.12. Polyethylene monolayers mounted on collodion before (*a*) and after (*b*) irradiation at 18 K in the electron microscope. Note the increase in lateral dimensions and the reversal of contrast from dark to light in comparison to the background. (From Grubb, Keller & Groves, 1972.)

oriented around *b*. From the above figures one would predict, following irradiation, a 23% increase in radius and 5% decrease in volume. To achieve this reduction in volume the film and the circumference must each shorten by 12%. If the film is to remain entire this can only be achieved by allowing the 'spherulite' to become non-planar and specifically a cone of semi-angle $\sin^{-1}(0.88/1.23) = 46°$. Cones of precisely this angle are indeed formed, of which Fig. 8.13 shows examples.

One can go further and consider the case of banded spherulites. It is generally the case that parallel-sided sections of polymer films are featureless when first brought into the electron microscope beam but rapidly develop contrast (Fig. 8.14). In banded spherulites one again finds rings (Fig. 8.15), corresponding to those seen in polarizing optics, but with the black and white contrast reversed between the two cases.

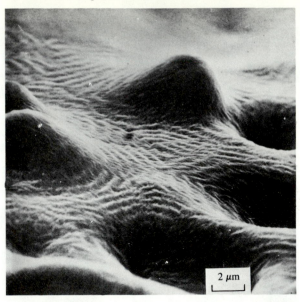

Fig. 8.13. A spherulitic film of polyethylene showing topography induced by irradiation in the scanning electron microscope. (From Grubb & Keller, 1972.)

What one now sees is contrast due to different mass thicknesses in alternate rings, with lamellae edge-on appearing darker. This is expected because of the expansions noted above. When lamellae are flat-on, c is parallel to the beam (so that in polarizing optics and also for diffraction contrast in the electron microscope, the tone would be dark). On irradiation, lamellae will contract by 40% in this direction, giving a lighter tone, especially by comparison with irradiated edge-on areas, where because a is vertical an expansion of 23% is expected. Provided the density remains uniform, this is already a sufficient explanation of the appearance of rings (and corresponds to detail in Fig. 8.15). It is due to mass transport of the same kind as measured in monolayers. In practice (e.g. Fig. 8.14) one also notices the splitting up of the image into lamellae, or lamellar packets, in regions where they are edge-on. This is entirely reasonable as the contraction down c will place lamellar interfaces in tension so that any yielding induced would produce the observed effect. There is thus a very satisfactory explanation of contrast effects in the electron microscopy of polymeric films. At the same time by showing how detail is created in the object, by what are essentially artefacts of the observation process, it provides a clear warning against incautious interpretation of microscopic images.

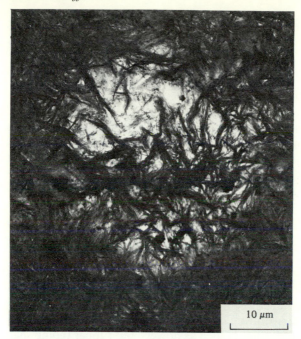

Fig. 8.14. The development of contrast within a 3 μm thick film of anabaric polyethylene in the 1 MV electron microscope. The light area was created within the focused beam of the microscope before taking the photograph. (From Bassett & Khalifa, 1976.)

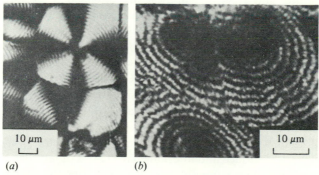

(a)   (b)

Fig. 8.15. Banding in different areas of the same polyethylene spherulites seen (a) optically between crossed polars and (b) by transmission electron microscopy of a thin section. (From Dlugosz & Keller, 1968.)

8.4   Further reading

The review by Grubb (1974) is recommended as a survey of radiation damage and the electron microscopy of organic polymers.

9   Aspects of mechanical behaviour

The aspects of the large and important subject of the mechanical properties of polymers which we consider are selected from those of particular
morphological interest and relevance.

9.1   Fracture

Polymers are largely synonymous with plastics. For the most part they
are indeed plastic, deforming substantially before failure, but there are
circumstances where this is not true. Most if not all polymers are brittle
sufficiently far below their glass transition temperatures, failing in tension at low strains. This is because then the reduction of molecular
mobility and consequent stiffening eventually remove the ability to
lower stress concentrations by local yielding so that cracks may propagate. Low-temperature brittleness is well illustrated by the way in which
familiar flexible objects such as flowers and rubber tubing shatter when
struck sharply after being cooled to liquid nitrogen temperatures
(~ 77 K). The same phenomenon is used to advantage in the microtomy
of polymer samples. If it is attempted to prepare sections at room
temperature, and especially if these are intended to be thin enough
($< \sim 20$ nm) for conventional transmission electron microscopy, they
are likely to be substantially deformed. Cooling to make samples stiffer
and brittle provides a remedy for this problem. It is usual to cool with
liquid nitrogen but for polyethylene useful gains are given with much
more modest temperatures (e.g. 250 K) which are readily attainable
using solid carbon dioxide. This is possibly because there are two glass
transition temperatures for polyethylene: 190 K is where the wholly
amorphous polymer (produced by drastic quenching of thin films)
crystallizes on warming while 260 K is suggested to relate to mobility in
the constrained surface and interlamellar regions of lamellae.

Brittleness at low temperatures also facilitates the preparation of
fracture surfaces for microscopic examination. A standard method is to
cleave cold specimens in liquid nitrogen with a razor blade and hammer.
The detail exposed is mostly genuine (there may be fracture-induced
striations parallel to the chain-axis, for example, which are artefacts) but
selective. Certain (thinner) populations fracture less cleanly than others
and so may be overlooked while particular planes may be exposed

preferentially. In melt-crystallized linear polyethylene there is a strong tendency to {200} which lie parallel to spherulitic radii with the result that important and diagnostic views down the radial *b* directions tend to be absent. On the other hand branched polyethylene usually breaks between the large basal surfaces of lamellae. Selectivity arises, in general, from the anisotropy of properties involved in fracture. On the simple Griffith's criterion, cracks can only propagate when the reduction in strain energy is enough to offset that required for the formation of two new surfaces. With both modulus and surface energy highly anisotropic certain planes are expected to be favoured over others. That in melt-crystallized linear polyethylene {200} should be exposed is interesting because it is thought that these are the planes of folding. There are examples where the fold plane does influence mechanical behaviour – e.g., cracking in solution-grown lamellae prefers to follow the growth planes. This is not invariably the case, however, and cracks do form in other directions. Moreover, the relative contribution of folds to the energy balance must be small because they are present in such small proportions. For anabaric polyethylene lamellae the amount may be as little as ~1% yet {200} remain the fracture planes. Its selection is, therefore, probably by the crystal structure rather than by the plane of folding *per se*. Widening the concept of fracture surface energy to include the work of fracture allows interlamellar regions to be more readily considered and it is fracture in these which predominates in branched polyethylenes. This is almost certainly due to the crack-blunting abilities of these enlarged regions in polymers of lower crystallinity and is not without practical significance in relation, for example, to stress-cracking resistance.

If samples are not brittle when they are broken surface detail may well be obscured by local yielding. This phenomenon can itself be revealing, particularly in the study of anabaric polyethylenes. For these the number of folds per molecule is often reduced to one or none (molecules of 10^5 or 5×10^4 mass in lamellae 0.5 μm thick) producing behaviour which brings out the molecular contribution to fracture and ductility in the context of a two-phase model of crystalline interiors plus surface and interlamellar regions. Consider, for example, the replication of fracture surfaces prior to electron microscopy. Usually, a two-stage process is adopted rather than the apparently simpler one-stage procedure of evaporating carbon on a surface followed by dissolution of the specimen. The reason is that melt-crystallized specimens tend to break the carbon (to which they are probably physically linked, as in detachment replication, Section 4.4) as they swell and distort in the solvent. Two-stage replication avoids this by first making an impression of the original surface, for example, in a film of cellulose acetate plasticized with acetone, which is itself replicated with carbon (and oblique metal shadowing for contrast) in the second stage. The first impression is recovered, when the film has dried, by stripping, i.e. pulling the sample away. As this is normally done at room temperature the polymer will usually be ductile, at least locally, and may deform in the operation leaving threads of deformed polymer attached to the replica. Pulled threads are frequently present on replicas for this reason† but their formation also depends critically on the degree of folding per molecule. This may be inferred when threads are only drawn from the thinner lamellae in a surface of anabaric polyethylene. It has been confirmed by showing that in anabaric lamellae of the same thickness there are more threads the higher the molecular mass – so much so that detail may be almost obliterated for masses of $\sim 10^6$. Conversely, crystallizing the same fraction into thinner lamellae, using lower temperatures and pressures, also produces more threads. The conclusion is that, for the same surface condition, ductility is greater the more molecules are folded. The qualification is necessary because folding is only secondary to the issue: it is the nature of interlamellar connection which is of primary importance. This is, however, related to the degree of folding because division of the molecular length by the crystal thickness gives the number of times a molecule has the possibility either of folding or forming a tie molecule. One may expect that in given crystallization conditions the ratio of the two probabilities will be constant. In general, therefore, the number of interlamellar links will be proportional to the number of folds. The constant of proportionality will vary with growth conditions being least for crystallization from dilute solution. Mats of such crystals are, of course, brittle despite the high degree of folding because of weak interlamellar linking. Anabaric linear polyethylenes are similarly brittle‡ until the molecular mass reaches $\sim 10^6$. Such samples can be drawn to $\sim \times 8$ at 80 °C. The ratio of molecular length to lamellar thickness ($\sim 0.5\ \mu m$) is then ~ 20, the same as for a molecule of 5×10^4 mass within lamellae 25 nm thick in typical atmospheric products. In these instances there is little doubt that the number of tie molecules is the relevant property promoting ductility. One needs to be cautious, however, in generalizing from this because it has been found that mats of branched polyethylene crystals filtered from solution are highly extensible (to $\sim \times 50$). This mode of preparation precludes an assembly of lamellae linked by tie

† For polyethylene they may be avoided to a large extent by stripping at -30 °C, though then the impression itself may be prone to crack.

‡ Except in tension along the c direction of fibre-oriented samples, presumably because this produces no tensile stress on the {200} fracture planes.

molecules. Instead, one presumably needs to think in terms of inter-penetration of the surface regions of adjacent lamellae produced during sedimentation. In branched materials, these regions are particularly extensive and are the cause of their relatively low crystallinity. It is likely that here, as in other instances, brittleness is avoided because of the efficiency of crack-blunting afforded by the enlarged surface regions.

The notorious brittleness of anabaric polyethylenes in general arises from a different but related cause, namely the segregation of low molecular mass populations. Molecular masses $< \sim 10^4$ need to be abstracted from linear polyethylenes before homogeneous products (judged, for example, by the melting endotherm) result from treatment at 5 kbar. Similar differentiation results from prolonged annealing of whole polyethylene *in vacuo* ($\gtrsim 9$ months at temperatures to 132 °C) with the same mechanical consequences. In both instances the segregated populations are inherently brittle with not more than about three folds per molecule and being physically concentrated they have become weak links leading to the failure of the entire sample. It has been shown that the low-melting populations are exposed following brittle failure and that failure does not occur under identical conditions when low mass molecules are removed in advance. More recent investigations reveal that the segregation leading to failure occurs in an organized pattern throughout the sample.

These experiments involve commercial linear polyethylenes which were cold drawn in sheet form to their natural draw ratios at atmospheric pressure. When annealed within 30 K of their melting point at 5 kbar, unconstrained sheets relaxed shortening as much as one third along their length and expanding in transverse directions. After treatment all samples containing molecules $< \sim 10^4$ in mass were brittle (others were not) and sheets of these samples broke at low annealing temperatures in attempts to hold them at constant length. All the brittle samples (but none of the others, including those from which the shortest molecules had been extracted) showed a pattern of flaws (Fig. 9.1). These are arranged along intersecting bands lying at $\sim 45°$ to the chain axis which themselves lie in planes parallel to the length and thickness of the sheet. They have the outline of tensile cracks parallel to the draw direction but are, in fact, filled with low-melting, low molecular mass polymer. This is confirmed by direct observation with a hot-stage microscope when their interiors melt some 6 K below the remainder of the sample. As might be expected the number, and in any one sample also the size, of the flaws increases with annealing temperature in the range where short molecules are expected to be molten. Comparing different polyethylenes, the quantity of segregated material in the flaws is in

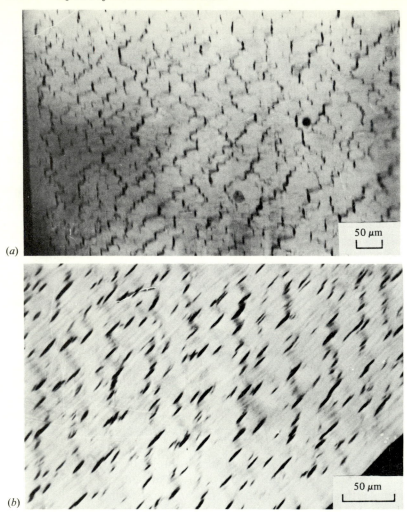

(a)

50 μm

(b)

50 μm

Fig. 9.1. Ordered arrangements of flaws (produced by annealing cold-drawn linear polyethylenes at 5 kbar) seen between crossed polars: (a) (tensile axis at 45 ° to the polars) fringes due to changing orientation link the flaws and illustrate the overall pattern; (b) a different sample oriented so that only flaws are in strong contrast; c at 45 ° to the vertical.

accord with the proportion of low molecular mass polymer. The detailed appearance of these regions may be seen in etched surfaces with the electron microscope (Fig. 9.2). They consist of thinner lamellae than the matrix with a tendency to lie parallel to the flaw boundary towards the outside but to lie parallel to the draw direction in the interior. Electron

(a)

(b)

Fig. 9.2. Detail of flaws in specimens like those in Fig. 9.1 viewed in the electron microscope after permanganic etching: (a) shows the bending of the matrix in relation to the pattern of flaws; and (b) detail of a flaw in which lamellae at the centre lie parallel to the tensile axis.

microscopy also confirms what is evident from polarizing microscopy, that the c axis of the matrix bends around the flaws, remaining more or less parallel to their surfaces.

The interpretation of this pattern begins with the observation that the flaws lie in lines at $\sim 45\,°$ to the draw direction. This is the expected direction of maximum resolved shear stress due to the contraction along c. That they should also be confined to planes containing the c axis which are perpendicular to the sheets suggests that the transverse stresses should be greater normal to the sheets rather than in their plane for this would locate the maximum shear stresses along the lines of flaws. The magnitude of the relative transverse expansions is indeed in this order. The phenomenon thus appears to be one of shear yielding which, however, only proceeds when there is molten polymer available to fill the cavities of the cracks thus nucleated. The molten polymer will crystallize on cooling (as thinner lamellae of the orthorhombic phase). The two orientations these possess are (i) the same chain direction as the surroundings at the edges and (ii) normal to the draw direction in the interiors, suggestive of nucleation on threads drawn across the width of the flaws. This segregated material is intrinsically brittle and when returned to room temperature samples fracture from flaw to flaw.

Spherulitic morphologies provide other examples of segregated textures which show brittleness in certain cases. The rejection of uncrystallizable or slowly crystallizing species will alter the mechanical properties of interfibrillar and interspherulitic boundaries. Only if the rejected species are short, however, will they promote brittleness. In stereospecific systems, such as isotactic polypropylene or polystyrene, atactic molecules will be among those removed to boundaries and, in principle, can be of any length. There will, nevertheless, be a general tendency to locate the shortest molecules between spherulites because these diffuse most rapidly and are best able to travel the distances involved. In consequence it may be expected in general that interspherulitic boundaries are more prone to brittleness than interfibrillar ones. The extent of the overall segregation will depend on the crystallization conditions, being greatest for slow growth at low supercoolings.

The facts for linear polyethylene are in general agreement with this picture. In this polymer it is the short molecules which segregate. At 130 °C they do this to such a degree that samples such as those of Figs. 4.27 and 4.35 are brittle even at room temperature. The breaks occur both within and between spherulites. More rapid growth at lower temperatures brings a change first to fracture occurring only in interspherulitic boundaries and eventually to ductile behaviour as discussed in Section 9.3. The general rule for minimizing textural brittleness is thus to try to reduce inhomogeneity by controlling the scale and extent of segregation through the growth rate and molecular mass distribution.

9.2   Crystal plasticity

One of the best understood aspects of the deformation of crystalline polymers is crystal plasticity whose active elements may be inferred from studies of diffraction patterns of deformed specimens. The behaviour revealed is in accord with the same principles which govern that for metals and other simpler solids. It is, however, necessary to consider also rubber-like forces attributable to a matrix within which crystals are embedded. The magnitude of these increases with the extent of deformation as molecular configurations become more distorted. For large deformations in tension (often $> \sim 200\%$) the original crystals are likely to have been disrupted. It is useful, therefore, to begin with effects concerning compression and relatively small strains (which can be applied either directly or through the agency of rolling) and, as a further simplification, to consider only transverse compressions, i.e. those perpendicular to the chain axis.

The first relevant experiments, concerning the development of texture in drawn then rolled branched polyethylene actually predate the discovery of polymer single crystals. These employed X-ray methods and established the basic features of interpretation which later works employing more precisely oriented samples and especially electron diffraction on stretched monolayers have confirmed and extended. There are three principal modes of deformation for polymer crystals: *slip*, *twinning* and *stress-induced phase (martensitic)* transformations. Each of these is discussed in turn, although in practice they often occur in association.

9.2.1   Deformation by slip

Slip involves relative movement across a *slip plane* (hkl) in the *slip direction* $[uvw]$ (Fig. 9.3); this is called the $[uvw]$ (hkl) mode. It is associated with the movement of dislocations, each of which, when gliding on the slip plane, contributes an increment of slip equal to its Burgers vector, also $[uvw]$ or a submultiple thereof. Those dislocations which are operative may be predicted on the basis of their strain energies which are proportional to the product of the shear modulus and the square of the Burgers vector. Because of the large elastic anisotropy of polymer crystals one must needs be cautious in comparing dislocations parallel and transverse to the chain direction on the basis of their Burgers vectors alone. In the transverse plane, however, moduli are the same or similar, depending on the crystal symmetry, so that predictions may confidently be made in this simple way by selecting the shortest lattice translations. In polyethylene, again the most studied material, the shortest lattice translation is b [010] followed by a [100]. It may also be expected that dislocations will be unable to move far if they would have to cut across covalently bonded chains. The slip (glide) planes should, therefore, be restricted to {hk0}.† A dislocation with b [010] as Burgers vector will then have (100) as slip plane while $a \langle 100 \rangle$ dislocations will glide on (010). It is well established from metallurgical studies that dislocations move once the resolved shear stress parallel to the slip direction reaches a critical value (the critical resolved shear stress). For a normal stress σ applied to a specimen the resolved shear stress along the slip direction is $\sigma \cos \theta \cos \chi$ where θ and χ are the respective angles between the applied stress, the normal to the slip plane and the slip direction (Fig. 9.3a). As θ and χ are just interchanged for [010] (100) and [100] (010) slip, the two modes will always experience the same driving stress. The former should thus always operate in preference to the latter because the resistance to dislocation motion is also roughly proportional to the strain energy, i.e. to the square of the Burgers vector.

The slip systems which have actually operated may be inferred from the orientation of planes and directions in a deformed specimen with respect to the tensile or compressive axis. Deformation in uniaxial tension moves the slip direction and the normal to slip plane continuously towards positions parallel and perpendicular to the tensile axis respectively‡ (Fig. 9.3b). In uniaxial compression, conversely, both slip plane and slip direction tend to become normal to the stress. Thus when predrawn branched polyethylene is subject to light rolling (equivalent to compression transverse to c) {100} planes align normal to the compression. This is evidence that slip of [010] (100) type has occurred.

9.2.2   Mechanical twinning

Mechanical twinning tends to occur as an alternative deformation mechanism to slip, especially in systems of low crystal symmetry and others whose limited number of slip systems makes it difficult to achieve a general change of shape. It is a common phenomenon in polyethylene which also shows a further parallel with metallic behaviour in that the

---

† In regularly folded solution-grown monolayers there is also a preference for the plane of folding.

‡ This assumes that only one slip system operates. When there are two symmetrically equivalent systems as, for example, in polyoxymethylene the final orientation has the two slip directions equally inclined to the stress axis, so that each is subjected to the same resolved shear stress, with their vector sum parallel to the tensile axis.

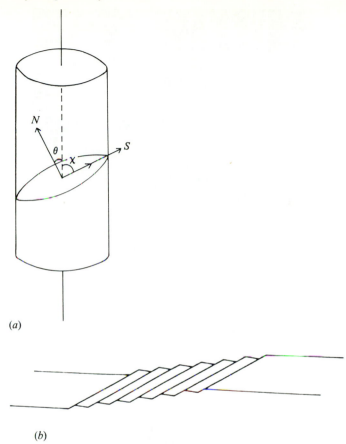

(a)

(b)

Fig. 9.3. (a) The slip plane with normal N and the slip direction S in a uniaxial test. (b) Slip rotates the slip plane and slip direction towards the tensile axis. (After Wyatt & Dew-Hughes, 1974.)

formation of kink bands, Fig. 9.4, (for example, by compression down c in fibre-oriented anabaric samples) produces discontinuities in the stress–strain curve owing to the higher stress required for nucleation than for propagation. Fig. 9.4 suggests rather clearly, as is generally true, that such transformations occur by homogeneous shear of the lattice parallel to the *composition plane* which is common to twin and parent structures. The twinning shear and atomic displacements are always small, in comparison to slip processes which may be illustrated by reference to the predominant {110} twinning mode of polyethylene, once the general terminology of twinning has been introduced.

Fig. 9.4. Kink band produced in fibre-oriented anabaric polyethylene by compression down *c*. (From Attenburrow & Bassett, 1979.)

Consider a sphere whose upper hemisphere undergoes uniform shear parallel to a direction η_1 lying in the composition plane K_1. K_1 itself is unchanged and there will be a second undistorted but rotated plane K_2 given by the intersection of the original sphere and the sheared hemisphere. These elements are drawn in Fig 9.5 together with the direction η_2 lying in K_2 which makes the same angle 2ϕ with η_1 after shear as it did with $-\eta_1$ before shear. The elements K_1, η_1, K_2 and η_2 are collectively known as the shear mode and define two reciprocal shears: on K_1 in the direction η_1 with K_2 as second undistorted plane or on K_2 in the direction η_2 with K_1 as second undistorted plane. The twinning shear *s*, defined as the displacement per unit distance away from the composition plane, is the same for both reciprocal shears being

$$s = 2 \cot 2\phi$$

The shear mode with smallest shear ($s = 0.25$) for polyethylene has either {110} or {310} as composition plane. The full elements are given in Table 9.1.

This is in accord with a generalization known as Mallard's law according to which symmetry planes lost in a lowering of symmetry (in this case {110} and {310} in a hexagonal structure on going to ortho-

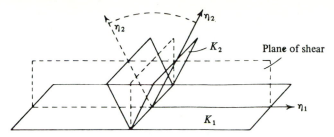

Fig. 9.5. The elements K_1, K_2, η_1 and η_2 of a shear mode. (From Wyatt & Dew-Hughes, 1974.)

Table 9.1. *Twinning elements in polyethylene*

Composition plane	Conjugate plane	Shear direction	Conjugate direction
K_1	K_2	η_1	η_2
(110)	(3$\bar{1}$0)	[1$\bar{1}$0]	[130]
(310)	(1$\bar{1}$0)	[1$\bar{3}$0]	[110]

rhombic) are likely to be composition planes for twins in the lower symmetry structure. What is actually involved in terms of atomic movements is shown in Fig. 9.6 for twinning on (110). Individual fold stems must translate parallel to the composition plane, in the direction [1$\bar{1}$0], and also rotate to bring the carbon–carbon zigzags into their correct positions and orientations in the twin.

The correct sense of shear stress to activate this particular shear may be obtained either by uniaxial tension near to *a* or uniaxial compression near to *b*. The specific {110} plane chosen will be that subject to the greatest resolved shear stress. In practice {110} twinning is found to result following strong rolling of predrawn polyethylene sheet. The rolling itself, being equivalent to a compressive stress, only aligns *a* parallel to the axis of compression as previously but in so doing it also generates elastic rubbery strains in the matrix. When the applied stress is removed the residual rubber-like stresses which are equivalent to compression along *b*, create the twinned orientation. The changes in diffraction pattern from which the change may be diagnosed are shown in Fig. 9.7. The reciprocal {310} shear is much less common, though why this should be so is not understood. There is little evidence of its occurring in stretched monolayers but its formation in bulk polyethylene is well documented and has led to the suggestion that the change from {110} to

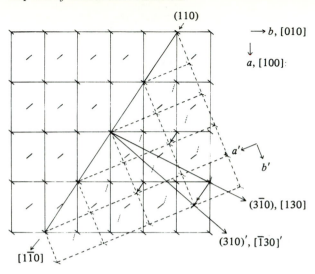

Fig. 9.6. Twinning on (110) in polyethylene, and the molecular movements necessary to restore the lattice.

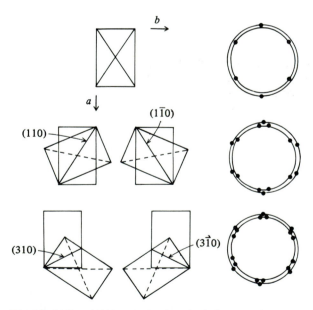

Fig. 9.7. {110} and {310} twinning in polyethylene and the resulting diffraction patterns (with the incident beam normal to the page). (From Bowden & Young, 1974a.)

{200} in the plane of folding may be the factor responsible for the difference in behaviour in the two morphologies.

9.2.3   Stress-induced phase transformation

Shear transformations are not restricted to reproduction of the parent lattice in twinned orientation. They may also lead to new crystal structures sharing a coherent, though now irrational, composition plane with the parent. These are known as martensitic transformations after the hard constituent of quench-hardened steel which is produced in this way. In polyethylene the structure which is generated is the monoclinic form (Fig. 9.8) which becomes stable if sufficient additional free enthalpy is provided by the product of stress and strain during transformation. In studies of stretched monolayers two orientations of the monoclinic cell have been found, with the *b* axes of the two structures making angles of 122° and 18°, in agreement with predictions for transformation modes called 1_1 and 2_1 respectively (Fig. 9.9). Mode 1 has the lowest possible shear strain (0.20) and mode 2 the next lowest

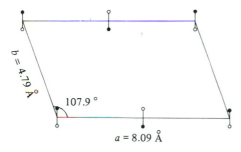

Fig. 9.8. The subcell structure of monoclinic polyethylene.

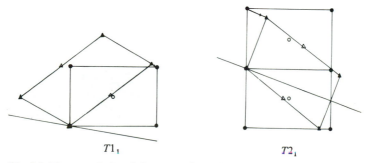

Fig. 9.9. The stress-induced shear transformations $T1_1$ and $T2_1$ for polyethylene with the composition planes between orthorhombic and monoclinic structures delineated. (After Bevis & Crellin, 1971.)

(0.32). Curiously, neither of the reciprocal shears (1_2 and 2_2) is found possibly because, it has been suggested, of the more awkward configurations of the composition planes. The two modes which are observed, 1_1 and 2_1, have for K_1 (1, 3.8, 0) and (1, 2.35, 0) respectively which are easily approximated by stepped (130) surfaces. The reciprocal shears, on the other hand, have composition planes of higher indices which should cause higher surface energies.

All these deformation mechanisms plus others of the same kind not specifically mentioned are in competition. Which one operates in given circumstances is likely to be determined by the resolved shear stress on the slip or twin (composition) plane exceeding the relevant critical value (~ 15 MN m^{-2} for each of the main slip and shear modes according to measurements on bulk polyethylene). In practice, it has been found for polyethylene monolayers stretched to 25% on a substrate with consequent imposed strains that the operative modes are predicted reasonably well according to the similar criterion of maximum resolved shear strain on the same planes. By using a population of true lozenges which collapsed only by shear (Section 4.2), making fold stems normal to the substrate, it has also been possible to distinguish the diffraction patterns originating from different sectors and so to differentiate between mechanisms applicable to different planes of folding. The differing modes for strains transverse to the chain axis are summarized in Fig. 9.10. Of particular interest is that $\langle 110 \rangle$ slip occurs, in addition to other mechanisms already described, but it operates over a much greater

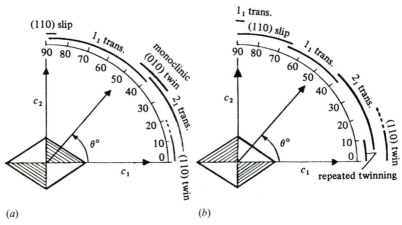

(a)   (b)

Fig. 9.10. Variation of deformation modes with the angle θ between the tensile axis and a for (a) (110) and (b) (1$\bar{1}$0) sectors in lozenge-shaped polyethylene monolayers. (From Allen & Bevis, 1974.)

angular range when the slip plane coincides with the fold plane (i.e. when glide would be between folded ribbons) than otherwise.

It is not yet possible to be as quantitative in the analysis of deformation modes in melt-crystallized polyethylene. Partly this is because of the spread of orientations in specimens. Even though this has been reduced to within $\pm 8\,^\circ$ of a single crystal texture in certain cases, this is sufficient to blur the distinction between competing modes. A more fundamental limitation is the inadequacy of our knowledge of interlamellar connexions and their consequences which could well, for example, invalidate the simplifying assumption that observed effects are equivalent to a collection of single crystals of the same range of orientations. The situation in practice is illustrated by Fig. 9.11 which refers to a polyethylene specimen in near single-crystal orientation compressed down b. At a strain of 37% both orthorhombic and monoclinic structures are present. The orthorhombic cell is, however, rotated $\sim 6\,^\circ$ more than for the expected {110} twinning; this has been attributed to an accompanying slip process but assignation of the exact mode has now become uncertain. The orientation between the orthorhombic and monoclinic cells

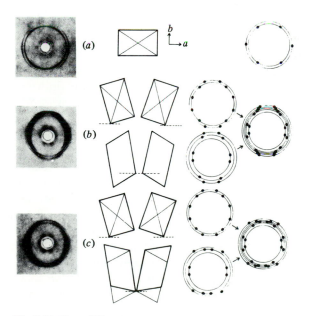

Fig. 9.11. X-ray diffraction patterns and their interpretation for a polyethylene specimen with near single-crystal orientation compressed parallel to b: (a) original, (b) under load at 37% strain and (c) after relaxation to 26% strain. (From Bowden & Young, 1974b.)

appears to be a more fixed quantity and now nearly coincides with that for the hitherto unrecorded Tl_2 mode: the angular separation from the Tl_1 position is, however, only $12°$. On the assumption that Tl_2 is operative in the deformation of bulk, but not monolayer, polyethylene it has been suggested that this, too, may be laid at the door of a change in fold plane. A very clear difference between the two cases is that on relaxation. Whereas monolayers stretched on a substrate do not change their diffraction patterns when the applied stress is released, this is not so for the bulk. In Fig. 9.11, when the strain has relaxed to 26%, the rubbery stresses have caused a marked change in orientation owing to twinning of the monoclinic phase.

9.2.4   Other modes

A number of other modes of deformation is encountered, particularly in more general conditions of deformation. Chief amongst these is [001] ($hk0$) slip, also known as chain slip or intralamellar shear. It is not an unreasonable inference that this occurs in the alignment of molecules along the tensile axis in drawn polymers although, as is discussed presently, crystals may be disrupted in the process. Other explanations are also possible, such as an affine deformation of crystalline units embedded in a rubbery matrix, which can account for increasing alignment with higher draw ratios in, for example, polyethylene terephthalate. There is no question, however, that intralamellar shear does occur widely: it is after all the shear mode with smallest Burgers vector in polyethylene. It may be recognized unambiguously in drawn PTFE and anabaric polyethylenes from sheared lamellae, with appropriately inclined c axis striations, in their fracture surfaces (Fig. 9.15).

Intralamellar shear is usually contrasted with interlamellar shear, in which neighbouring lamellae are undistorted but slide past each other along their adjacent fold surfaces. The latter shear is believed to be prominent in the deformation of polyethylene near its melting point. These and other modes involving deformation of the interlamellar 'amorphous' material need to be investigated by a combination of low-angle and wide-angle X-ray diffraction among other techniques. Fig. 9.12 shows how the two types of pattern are complementary and that, in this example, the relaxation on annealing of drawn then rolled branched polyethylene, they are consistent with interlamellar shear. In this context lamellae are rotated to lie normal to the axis of compression by rubbery stresses. An alternative suggestion is that lamellae may rotate together as stacks. In fact all of these modes occur in practice. Direct observation of tensile deformation of anabaric polyethylenes shows intralamellar and interlamellar shear, lamellar separation and

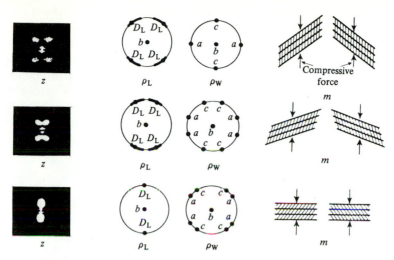

Fig. 9.12. Small-angle X-ray patterns, with pole figures both for these (ρ_L) and corresponding wide-angle patterns (ρ_W) plus their interpretation in terms of lamellae (*m*). These represent a stage of the relaxation on heating of a drawn then rolled sheet of branched polyethylene. (From Hay & Keller, 1967.)

rotation, all mechanisms which had been inferred to operate in the deformation of chainfolded material from diffraction behaviour.

9.3   On drawing and drawn fibres

Tensile deformation or drawing is the predominant means of producing polymeric fibres which are so important commercially: others are extrusion of solid or melt and methods linked to flow crystallization (Section 7.3). Morphologically, however, both the nature of the product and the processes involved in its formation can hardly be accurately described. The best that can yet be done is to synthesize various aspects, or apparent aspects, into a coherent model and compare this with experiment. There are such models but none that commands wide acceptance and is without controversy. This situation is likely to persist until more and better morphological information, especially electron microscopy to establish the spatial structures, is forthcoming. Fortunately, this appears imminent and significant progress has begun to be made.

The phenomena of drawing represent a series of variations upon a theme, or rather several themes, for there are so many relevant variables, including the temperature of drawing, its strain rate, initial morphology and molecular mass distribution, among others. Stretching at room temperature is known as cold draw and obviously concerns deformation of pre-existing structures. The manner of deformation alters as one proceeds to hot draw, at higher temperatures, and the relative critical stresses to activate various modes change. Eventually, one finishes up with combined crystallization and deformation when molten filament from a spinneret is wound up, as happens commercially, probably leading to the formation and disruption of row structures (Fig. 2.7a). So vast is this canvas that here only the broadest outlines of relevant matters can be sketched in.

9.3.1   Deformation of spherulites

A feeling for the processes involved in drawing may be obtained by stretching thin spherulitic films and observing them optically between crossed polars plus (with more difficulty) X-ray microdiffraction. It is convenient to prepare films with banded spherulites of polyethylene by slowly cooling a solution in o-chlorophenol and removing the skin of polymer formed on the surface. If these retain solvent, or have it replaced by microscope immersion oil, they are ductile and may be stretched by as much as 10 times.

The behaviour on stretching is both elastic and plastic, with both aspects tending to occur under the same conditions in different parts of the same film. Elastic behaviour persists until about fivefold extension and consists of homogeneous deformation in which spherulites retain their banding but become elliptical (Fig. 9.13a). Remarkably, if the stress is released, if necessary with help from a rise in temperature, homogeneously deformed spherulites will relax under the action of rubbery stresses and resume their original appearance. In contrast, plastic behaviour is inhomogeneous, though occurring systematically at specific locations within the morphology. In some cases interspherulitic boundaries yield (Fig. 9.13b) producing zones transverse to the draw direction. This is believed to be a consequence of segregation-induced softening because of observations that this ductility is possessed by those spherulites which are grown just fast enought to escape brittleness (Section 9.1). Faster growth still places further limits on segregation and boundary yielding disappears.

The usual plastic behaviour results from yielding along that diameter of a spherulite which is normal to the draw direction. This is expected to be the weakest linkage because the entire stress then falls on each radial

Fig. 9.13. Stages in the deformation of banded polyethylene spherulites: (a) homogeneous and quasi-elastic; (b) yielding at the boundary between spherulites; and (c) highly drawn with voids and residual caps of comparatively undeformed spherulite. (From Hay & Keller, 1965.)

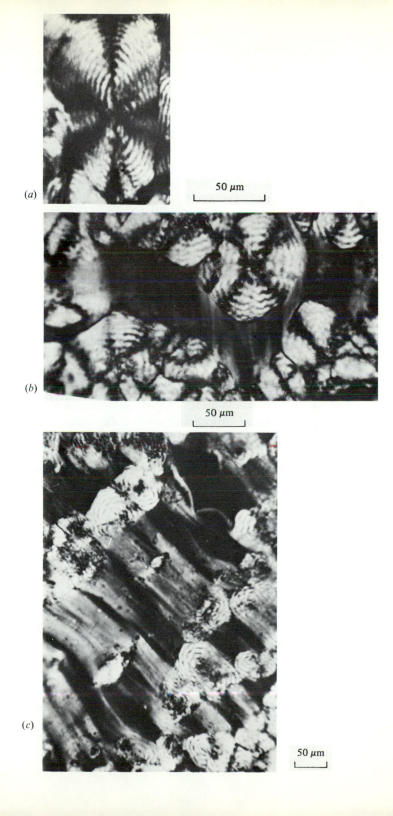

(a)

50 μm

(b)

50 μm

(c)

50 μm

element in series. In other orientations, adjacent units are able to reinforce each other to increasing extents as their angle to the draw direction decreases. The plastic zone expands until, as in Fig. 9.13*c*, only two opposing caps of seemingly undeformed and still banded spherulite remain, surrounded by highly extended material which retains little or no memory of the original morphology. Eventually, a fibrous texture is attained (which is opaque because it scatters light) within which longitudinal cracks tend to develop along the remains of what were once interspherulitic boundaries elongated along the draw direction.

The change from a texture clearly related to the original to one seemingly independent of it appears to occur gradually for the specimen as a whole although it is claimed that it occurs abruptly for individual lamellae which are supposed to be disrupted by micronecking into microfibrils (1–3 nm wide) and then deformed further. Certainly in bulk specimens it accompanies necking, i.e. a rapid decrease in dimensions of a drawn sample. This is a common, but not universal, feature in the drawing of polymers which can also occur in metals, for example. It occurs at a particular point on the stress–strain curve when further extension fails to produce an additional force. A specimen may then contrive to lengthen until the extended material work-hardens sufficiently to contain the stress.

Examination of the various morphologies mentioned with the electron microscope is difficult – more so than for the starting materials because of the greater subtlety of texture. In anabaric polyethylenes (of molecular mass $> \sim 10^6$) it has, however, been possible to follow changes to eightfold extension (which may be achieved by drawing at 80 °C) using permanganic etching. This is a particularly valuable system to study, principally because if lamellae are suitably thick they can only have been formed at high pressure. In this system, unlike others, a clear distinction is thus possible between lamellae which were present before deformation (i.e. those ~ 0.5 μm thick) and those which have formed subsequently (~ 20 nm thick).

The orientation in such samples undergoes similar changes to those of spherulitic films (drawn to extensions to about six): 200 poles concentrate transverse to the draw while 110 and 020 are split but move continuously towards this position for greater extensions. At the same time the melting endotherm, which initially indicated only high melting material, develops increasing proportions of low melting polymer (Fig. 9.14). The corresponding lamellar morphology is shown in Fig. 9.15. It consists of original lamellae which have become highly sheared, so that their chain axes and their planes approach, but have yet to reach, the draw direction. They have also increased their aspect ratio by lengthen-

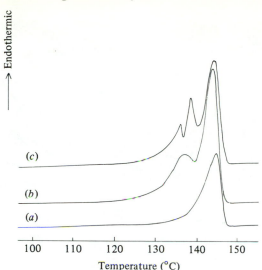

Fig. 9.14. Melting endotherms of anabaric high molecular mass polyethylene: (*a*) original, (*b*) after drawing fivefold at 80 °C and (*c*) as (*b*) but after annealing at constant length and 130 °C.

ing (~ 1.5 times) along the draw and approximately halving their width in the transverse direction. Surrounding these lamellae is a matrix of reformed polymer, now with lamellae ~ 20 nm thick which must have come from disruption of thick lamellae during deformation. Its location is particularly clearly revealed in Fig. 9.15*b* showing the result of annealing at 130 °C. This treatment has caused the matrix to recrystallize with a chevron pattern pointing along the draw direction. Of particular interest is the survival of thick lamellae at eightfold extension not only for its own sake but because the sample cannot have been molten: such thick lamellae form only at high pressure and melting would have led to eventual crystallization at ~ 20 nm thickness.

It is often supposed that melting is involved in reforming the structure especially when it passes through a neck. The evidence against this hypothesis is, however, very strong. In addition to that just cited it includes measurement of the energy available in the deformation which is insufficient and the fact that necking may proceed at very slow rates (i.e. isothermal conditions). Nevertheless, there remains the likelihood of local temperature rise during deformation which must be related, in some way still to be clarified, to the fact that the long period in a highly deformed polymer (i.e. beyond the neck or when molecules are aligned parallel to draw) depends only upon the temperature of drawing and not

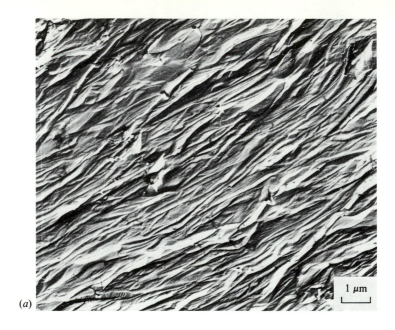

(a)

1 μm

(b)

1 μm

Fig. 9.15.

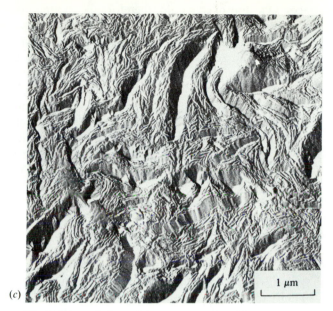

(c)

Fig. 9.15. Morphology of the drawn samples of Fig. 9.14 revealed by permanganic etching: (*a*) drawn; (*b*) and (*c*) drawn and annealed. The tensile axis is parallel to the page in parts (*a*) and (*b*) and perpendicular to it in part (*c*).

upon the draw ratio or the initial morphology. Although the precise manner of its formation still requires substantiation the type of crystalline arrangement within the highly drawn region is becoming apparent. A discussion of this follows in the context of ultra-high-modulus fibres.

9.3.2   The Young's modulus of drawn fibres

The existence of polymer lamellae, especially chainfolded lamellae, gives an immediate qualitative understanding of why the Young's modulus of macroscopic samples is typically 3–4 orders of magnitude less than values for the crystal lattice. It is because measured values (of a series arrangement of crystalline lamellae alternating with soft interlamellar regions) relate primarily to the low modulus disordered material. If the modulus of the crystalline lattice could be approached in practice it would, however, bring a new range of potential applications to polymers because these values can exceed the Young's modulus of steel $(210 \, \text{GN m}^{-2})$.

What needs to be done is to maximize the potential modulus by having the greatest number of polymer chains per unit area (i.e. by using polyethylene), to align all chains parallel and to mitigate the weakness of the fold surfaces. One approach would, therefore, be to prepare extended-chain polyethylene. Anabaric materials have high chain-extension (typically ~ 25 times greater than for polyethylene crystallized at 1 bar) and, by pressure annealing cold-drawn starting material, may be prepared with a common chain orientation. They show, however, only a small increase in modulus because their wide interfaces, although much reduced in number per unit volume, remain soft and confer this property upon what is still essentially a series arrangement of lamellae. Moreover, the extra stiffening incorporated in the starting material by cold-drawing is removed during pressure annealing, showing that the ordering given by crystallization or annealing treatments tends to lower Young's modulus.

The crucial factor is, therefore, to stiffen the soft lamellar interfaces with extended sequences of covalent bonds (taut tie molecules in some descriptions). As the modulus of drawn fibres is enhanced, increasingly so with draw ratio, such bridging has long been thought to be responsible in principle. Nevertheless, the modulus of conventional drawn fibres is still typically two orders of magnitude lower than the theoretical figure of ~ 300 GN m^{-2} (the figure varies slightly according to different estimates). What have recently been found, initially by Capaccio & Ward (1973), are empirical ways – based on control of molecular mass distribution and starting morphology – to achieve ultra-high draw ratios of ~ 30. The Young's modulus continues to increase with draw ratio and by this extension has reached ~ 70 GN m^{-2}; still higher figures have been claimed.

A microscopic description of these materials is also emerging, with general application to drawn fibres. Nitric acid is no longer able to penetrate and degrade polyethylene fibres quickly at draw ratios $> \sim 20$ but surfaces etched with this reagent show periodicities of ~ 20 nm, invariant with the extension at a given draw temperature, in agreement with the long period data. At the same time the melting endotherm shows evidence of superheating, i.e. circumstances in which lamellae are unable to melt sufficiently quickly to avoid being temporarily heated above their equilibrium melting points. This is a familiar feature of virgin PTFE or anabaric polyethylenes and is usually associated with highly constrained molecular conformations which are unable to adjust rapidly enough to the random molten state. Furthermore, the line width of the 002 reflection, supported by electron microscopic dark-field microscopy, shows coherent crystalline lengths in the chain direction greater than the long period and increasing with draw ratio from ~ 20 nm to ~ 50 nm. A morphological picture capable of accounting in a straightforward way for all these observations is of small individual crystallites linked by increasing numbers and lengths of intercrystalline bridges (sufficiently wide to give coherent X-ray diffraction) as the draw ratio lengthens overall molecular extension (Fig. 9.16). Moreover, the degree of linking may be measured as the ratio of the coherent length to

Fig. 9.16. Schematic representation of the structure of the crystalline phase of highly oriented linear polyethylene. (From Gibson, Davies & Ward, 1978.)

X-ray long period and used to correlate the observed moduli in a quantitative manner. This is probably the best example of practical gains which have yet been attained by manipulation of polymer morphology. At the same time it hints at how much more morphological understanding is necessary to complete the picture and optimize the rewards. The combination of achievement and promise fairly reflects the present-day character of the field of polymer morphology.

9.4   Further reading

This chapter has considered just a few mechanical topics of morphological interest. Standard texts on the field of mechanical properties of polymers are Andrews (1968) and Ward (1971). In addition, the review by Bowden & Young (1974*a*) is particularly helpful in the consideration of yield and its crystallography.

References

Allan, P., & Bevis, M. (1974) *Proc. Roy. Soc. A*, **341**, 75.

Andrews, E.H. (1968) *Fracture in Polymers*. London: Oliver & Boyd.

Attenburrow, G.E., & Bassett, D.C. (1979) *J. Mat. Sci.*, **14**, 2679.

Balta Calleja, F.J., Bassett, D.C., & Keller, A. (1963) *Polymer*, **4**, 269.

Barnes, J.D., & Khoury, F. (1974) *J. Res. Nat. Bur. Std.*, **78A**, 363.

Barnett, J.D., Block, S., & Piermarini, G.J. (1973) *Rev. Sci. Inst.*, **44**, 1.

Bassett, D.C. (1961) *Phil. Mag.*, **6**, 1053.

Bassett, D.C. (1964a) *Polymer*, **5**, 457.

Bassett, D.C. (1964b) *Phil. Mag.*, **10**, 595.

Bassett, D.C. (1965) *Phil. Mag.*, **12**, 907.

Bassett, D.C. (1968a) *Phil. Mag.*, **17**, 145.

Bassett, D.C. (1968b) *J. Cryst. Growth*, **3, 4**, 761.

Bassett, D.C. (1976) *Polymer*, **17**, 460.

Bassett, D.C. (1977) *High Temp. High Press.*, **9**, 553.

Bassett, D.C., Block, S., & Piermarini, G.J. (1974) *J. Appl. Phys.*, **45**, 4146.

Bassett, D.C., & Carder, D.R. (1973) *Phil. Mag.*, **28**, 513.

Bassett, D.C., Dammont, F.R., & Salovey, R. (1964) *Polymer*, **5**, 579.

Bassett, D.C., Frank, F.C., & Keller, A. (1959) *Nature*, **811**, 810.

Bassett, D.C., Frank, F.C., & Keller, A. (1963a) *Phil. Mag.*, **8**, 1739.

Bassett, D.C., Frank, F.C., & Keller, A. (1963b) *Phil. Mag.*, **8**, 1753.

Bassett, D.C., & Hodge, A.M. (1978a) *Polymer*, **19**, 469.

Bassett, D.C., & Hodge, A.M. (1978b) *Proc. Roy. Soc. A*, **359**, 121.

Bassett, D.C., Hodge, A.M., & Olley, R.H. (1979) *Disc. Faraday Soc.*, **68**, 218.

Bassett, D.C., & Keller, A. (1962) *Phil. Mag.*, **7**, 1553.

Bassett, D.C., Keller, A., & Mitsuhashi, S. (1963) *J. Polymer Sci. A*, **1**, 73.

Bassett, D.C., & Khalifa, B.A. (1976) *Polymer*, **17**, 275.

Bassett, D.C., Khalifa, B.A., & Olley, R.H. (1976) *Polymer*, **17**, 286.

Bassett, D.C., Khalifa, B.A., & Olley, R.H. (1977) *J. Polymer Sci. (Phys. Edn.)*, **15**, 995.

Bassett, D.C., & Turner, B. (1972) *Nature (Phys. Sci.)*, **240**, 146.

Bassett, D.C., & Turner, B. (1974a) *Phil. Mag.*, **29**, 285.

Bassett, D.C., & Turner, B. (1974b) *Phil. Mag.*, **29**, 925.

Bevis, M., & Crellin, E.B. (1971) *Polymer*, **12**, 666.

Blackadder, D.A., & Lewell, P.A. (1970a) *Polymer*, **11**, 125.

Blackadder, D.A., & Lewell, P.A. (1970b) *Polymer*, **11**, 147.

Blackadder, D.A., & Lewell, P.A. (1970c) *Polymer*, **11**, 659.

Bowden, P.B., & Young, R.J. (1974a) *J. Mat. Sci.*, **9**, 2034.

Bowden, P.B., & Young, R.J. (1974b) *Phil. Mag.*, **29**, 1061.

Bunn, C.W. (1939) *Trans. Faraday Soc.*, **35**, 482.

Bunn, C.W. (1953) *Fibres from Synthetic Polymers*, (Hill, R. ed.), pp. 240–300. Amsterdam: Elsevier.

Capaccio, G., & Ward, I.M. (1973) *Nature (Phys. Sci.)*, **243**, 143.

Daubeny, R. de P., Bunn, C.W., & Brown, C.J. (1954) *Proc. Roy Soc. A*, **226**, 531.

Dlugosz, J., & Keller, A. (1968) *J. Appl. Phys.*, **39**, 5776.

Fischer, E.W. & Schmidt, G.F. (1962) *Angew. Chem.*, **74**, 551.

Fitchmun, D.R. & Newman, S. (1970) *J. Polymer Sci. A-2*, **8**, 1543.

Frank, F.C. (1964) *Proc. Roy. Soc. A*, **282**, 9.

Frank, F.C., & Tosi, M.P. (1961) *Proc. Roy. Soc. A*, **263**, 323.

Geil, P.H. (1963) *Polymer Single Crystals*, Chichester: Wiley Interscience.

Geil, P.H., Anderson, F.R., Wunderlich, B., & Arakawa, T. (1964) *J. Polymer Sci. A*, **2**, 3707.

Gibson, A.G., Davies, G.R., & Ward, I.M. (1978) *Polymer*, **19**, 683.

Girolamo, M., Keller, A., Miyasaka, K., & Overbergh, N. (1976) *J. Polymer Sci. (Phys. Edn.)*, **14**, 39.

Grubb, D.T. (1974) *J. Mat. Sci.*, **9**, 1715.

Grubb, D.T., & Keller, A. (1972) *J. Mat. Sci.*, **7**, 822.

Grubb, D.T., Keller, A., & Groves, G.W. (1972) *J. Mat. Sci.*, **7**, 131.

Harrison, I.R., & Baer, E. (1971) *J. Polymer Sci. A-2*, **9**, 1305.

Hay, I.L., & Keller, A. (1965) *Kolloid Z.*, **204**, 43.

Hay, I.L., & Keller, A. (1967) *J. Mat. Sci.*, **2**, 538.

Hodge, A.M., & Bassett, D.C. (1977) *J. Mat. Sci.*, **12**, 2065.

Hoffman, J.D. (1979) *Polymer*, **20**, 1071.

Hoffman, J.D., Davis, G.T., & Lauritzen, J.I. (1976) in *Treatise on Solid State Chemistry*, vol. 3 (Hannay N.B. ed.), chap. 7. New York: Plenum.

Keith, H.D. (1963) in *Physics and Chemistry of the Organic Solid State*, (Fox, D. Labes, M.M. & Weissberger, A. eds.), Chichester: Wiley Interscience.

Keith, H.D. (1964*a*) *J. Polymer Sci. A*, **2**, 4339.

Keith, H.D. (1964*b*) *J. Appl. Phys.*, **11**, 3115.

Keith, H.D., & Padden, F.J. (1959*a*) *J. Polymer Sci.*, **39**, 101.

Keith, H.D., & Padden, F.J. (1959*b*) *J. Polymer Sci.*, **39**, 123.

Keith, H.D., & Padden, F.J. (1963) *J. Appl. Phys.*, **34**, 2409.

Keith, H.D., & Padden, F.J. (1964*a*) *J. Appl. Phys.*, **35**, 1270.

Keith, H.D., & Padden, F.J. (1964*b*) *J. Appl. Phys.*, **35**, 1286.

Keith, H.D., Padden, F.J., & Vadimsky, R.G. (1966) *J. Polymer Sci. A-2*, **4**, 267.

Keith, H.D., Padden, F.J. & Vadimsky, R.G. (1971) *J. Appl. Phys.*, **42**, 4585.

Keller, A. (1968) *Rep. Prog. Phys.*, **31**, 623.

Keller, A. (1958) in *Growth and Perfection of Crystals*, (Doremus, R.H. Roberts, B.W. & Turnbull, D. eds.), p. 499. Chichester: Wiley Interscience.

Keller, A. (1959*a*) *Makromol. Chem.*, **34**, 1.

Keller, A. (1959*b*) *J. Polymer Sci.*, **39**, 151.

Keller, A. (1967) *Kolloid Z.*, **219**, 118.

Keller, A. (1968) *Rep. Prog. Phys.*, **31**, 623.

Keller, A., & Bassett, D.C. (1960) *J.R. Micr. Soc.*, **49**, 243.

Keller, A., & Pedemonte, E. (1973) *J. Crystal Growth*, **118**, 111.

Khoury, F. (1966). *J. Res. Nat. Bur. Std.*, **70A**, 29.

Khoury, F., & Barnes, J.D. (1972) *J. Res. Nat. Bur. Std.* **76A**, 225.

Khoury, F., & Barnes, J.D. (1974) *J. Res. Nat. Bur. Std.*, **78A**, 95.

Khoury, F., & Passaglia, E. (1976) in *Treatise on Solid State Chemistry*, vol. 3, (Hannay, N.B. ed.), chap. 6. New York: Plenum.

Kitaigorodskii, A.I. (1961) *Organic Chemical Crystallography*, pp. 177–215. New York: Consultants Bureau.

Kovacs, A.J., Gonthier, A., & Straupe, C. (1975) *J. Polymer Sci. C*, **50**, 283.

Lauritzen, J.I., & Hoffman, J.D. (1960) *J. Res. Nat. Bur. Std.*, **64A**, 73.

Lauritzen J.I., di Marzio, E.A., & Passaglia, E. (1966) *J. Chem. Phys.*, **45**, 4444.

Lauritzen, J.I., & Passaglia, E. (1967) *J. Res. Nat. Bur. Std.*, **71A**, 261.

McCrum, N.G., & Lowell, P.N. (1967) *J. Polymer Sci. B*, **5**, 1145.

Mackley, M.R., & Keller, A. (1975) *Phil. Trans.*, **278**, 29.

Mandelkern, L. (1964) *Crystallization of Polymers*. New York: McGraw–Hill.

Mandelkern, L. (1968) *J. Mat. Sci. B*, **6**, 615.

Matsuoka, S. (1962) *J. Polymer Sci.*, **57**, 569.

Maxwell, B. (1965) *J. Polymer Sci. C*, **9**, 43.

Mitsuhashi, S., & Keller, A. (1961) *Polymer*, **2**, 109.

Natta, G., & Corradini, P. (1960) *Nuovo Chim.*, *Suppl. to vol. 15*, **1**, 40.

Olley, R.H., Hodge, A.M., & Bassett, D.C. (1979) *J. Polymer Sci. (Phys. Edn.)*, **17**, 627.

Palmer, R.P., & Cobbold, A.J. (1964) *Makromol. Chem.*, **74**, 174.

Pennings, A.J. (1977) *J. Polymer Sci., (Symposia)*, **59**, 55.

Peterman, J., & Gleiter, H. (1976) *J. Polymer Sci. (Phys. Edn.)*, **14**, 555.

Price, F.P. (1959) *J. Polymer Sci.*, **39**, 139.

Rees, D.V., & Bassett, D.C. (1971) *J. Mat. Sci.*, **6**, 1021.

Reneker, D.H. (1962) *J. Mat. Sci.*, **59**, 539.

Sadler, D.M. (1968) Ph.D. Thesis, University of Bristol.

Sadler, D.M. (1971) *J. Polymer Sci. A-2*, **9**, 779.

Salovey, R., & Bassett, D.C. (1964) *J. Appl. Phys.* **35**, 3216.

Salovey, R., & Keller, A. (1961a) *Bell Syst. Tech. J.*, **40**, 1397.

Salovey, R., & Keller, A. (1961b) *Bell Syst. Tech. J.*, **40**, 1409.

Sanchez, I.C. (1974) *J. Macromol. Sci., Rev. Macromol. Chem.*, **10**, 113.

Schultz, J.M. (1974) *Polymer Materials Science*. Englewood Cliffs, NJ: Prentice Hall.

Statton, W.O., & Geil, P.H. (1960) *J. Appl. Polymer Sci.*, **32**, 2332.

Stein, R.S., & Rhodes, M.B. (1960) *J. Appl. Phys.*, **31**, 1873.

Tadokoro, H. (1979) *Structure of Crystalline Polymers*. Chichester: Wiley Interscience.

Ward, I.M. (1971) *Mechanical Properties of Solid Polymers*. Chichester: Wiley Interscience.

Williams, T., Blundell, D.J., Keller, A., & Ward, I.M. (1968) *J. Polymer Sci. A–2*, **6**, 1613.

Wittman, J.C., & Kovacs, A.J. (1970) *Berichte der Bunsengesellschaft für Physikalische Chemie*, **74**, 901.

Wunderlich, B. (1973) *J. Polymer Sci. C*, **43**, 29.

Wyatt, O.H., & Dew-Hughes, D. (1974). *Metals, Ceramics and Polymers*. Cambridge University Press.

Index